Sustainable Materials and Technology

Sustainable Materials and Technology (SMT) book series publishes research monographs (both edited and authored volumes) showcasing the latest developments in the field and comprehensively covering topics such as:

- Recycling of waste into useful material and their energy applications
- Catalytic action of Nano oxides for efficient carbon reforming process
- Sustainable technologies for plastic transformation
- Bifunctional nanoparticles for sustainable water splitting applications
- Sustainable dying and printing
- New materials from waste
- Sustainable Manure Management and Technology: Potentials, Uses and limitations
- Sustainable Mechanical Engineering Approach
- Sustainable biochemistry for the improvement of health
- Sustainable development of Mechanical recycling of automotive components
- Sustainable-waste recycling and conversion in useful materials for different applications
- Sustainable development of inexpensive Nano-photo catalysts
- Sustainable development of recycling of discarded lithium ion batteries
- Modern sustainable cement and concrete
- Sustainable adsorbent for hazardous removal
- Sustainable superior electromagnetic shielding materials
- Excellent sustainable nanostructured materials for energy storage device
- Sustainable development of heavy metal detoxification from water
- Carbon dioxide utilization for sustainable energy
- Sustainable development in green syntheses of materials
- Environment friendly and sustainable cloth for garments application
- Sustainable design and application of eco-materials
- Nanoparticles for sustainable environment applications
- Sustainable remediation of industrial contaminated water towards potential industrial applications
- Biomaterials for sustainable bioremediations

Anish Khan · Mohammad Jawaid ·
Khalid A. Alzahrani
Editors

Uses and Products of Recycled Expanded Polystyrene Foam Wastes

Springer

Editors
Anish Khan
Centre of Excellence for Advanced Materials Research
King Abdulaziz University
Jeddah, Saudi Arabia

Mohammad Jawaid
Department of Chemical and Petroleum Engineering, College of Engineering
United Arab Emirates University
Al Ain, United Arab Emirates

Khalid A. Alzahrani
Department of Chemistry, Faculty of Science
King Abdulaziz University
Jeddah, Saudi Arabia

ISSN 2731-0426 ISSN 2731-0434 (electronic)
Sustainable Materials and Technology
ISBN 978-981-96-5617-2 ISBN 978-981-96-5618-9 (eBook)
https://doi.org/10.1007/978-981-96-5618-9

This Springer imprint is published by the registered company Springer Nature Singapore Pte Ltd.
The registered company address is: 152 Beach Road, #21-01/04 Gateway East, Singapore 189721, Singapore

Contents

Mechanical Recycling of Expanded Polystyrene Foams

Maisha Farzana Antora, Hassan Tawsif Tazwar, and Adib Bin Rashid

Abstract Expanded polystyrene (EPS) foam, a widely used material for packaging and insulation, presents significant environmental challenges due to its non-biodegradable nature. This chapter explores the mechanical recycling of EPS, emphasizing its importance and benefits. It covers the entire recycling process, starting with collecting and sorting EPS waste from industrial and consumer sources. Pre-processing steps, including cleaning, contaminant removal, and size reduction, are detailed to prepare the material for recycling. The core processes of mechanical recycling densification, pelletizing, and extrusion are examined, highlighting the equipment and techniques used to transform EPS waste into reusable material. Quality control measures essential for maintaining recycled EPS's integrity, including physical and chemical property assessments and compliance with relevant standards, are discussed. Applications of recycled EPS for traditional uses like packaging and insulation and innovative applications in construction and consumer products are presented. The chapter addresses challenges such as contamination and economic viability, exploring technological advances and emerging research to overcome these issues. Through case studies and future directions, the chapter underscores the need for continued innovation and stakeholder engagement to enhance EPS recycling and foster a circular economy.

Keywords EPS foam · Mechanical recycling · Waste management · Circular economy

---

M. F. Antora · H. T. Tazwar
Department of Aeronautical Engineering, Military Institute of Science and Technology (MIST), Dhaka, Bangladesh

A. B. Rashid (✉)
Department of Mechanical Engineering, Military Institute of Science and Technology (MIST), Dhaka, Bangladesh
e-mail: adib8809@gmail.com

A. Khan et al. (eds.), *Uses and Products of Recycled Expanded Polystyrene Foam Wastes*, Sustainable Materials and Technology,
https://doi.org/10.1007/978-981-96-5618-9_1

1 Introduction

1.1 Overview of Expanded Polystyrene (EPS) Foam

Expanded polystyrene is a lightweight, low-density rigid plastic product utilized for insulation and packaging, among other uses. It is commonly known as "Styrofoam". It is a closed-cell foam material derived from styrene and pentane, both of which are hydrocarbons sourced from petroleum and natural gas byproducts. Styrene forms the cellular structure, while pentane acts as a blowing agent [1]. It primarily consists of air, with a small amount of polystyrene. The air trapped within EPS cells contributes significantly to its insulation properties. Unlike other foams, the air remains trapped and does not degrade over time. This ensures consistent insulation performance throughout the product's lifespan.

Grey or silver EPS products incorporate carbon or graphite, which reduces radiant heat transfer and improves thermal insulation [2]. EPS is characterized by many good properties such as low density, heat-insulating property, moisture resistance, durability, sound-absorbing property and low coefficient of heat conductivity. Many researchers reported that the EPS's static strength and Young's modulus depend on density. High-density EPS has also been found to possess greater energy absorption capacity than low-density ones. It can also be observed that the compressive strength increases with the strain rate when the strain rate is over 113 1/s, whereas the densification strain decreases with the strain rate [3]. The desirable characteristics of EPS, a lightweight, rigid material, is makes it well-suited for applications where weight reduction is paramount. Its robust structure offers superior support and protection. Furthermore, EPS possesses exceptional insulating properties, effectively retaining or repelling heat as required. Additionally, its ability to absorb shocks and vibrations safeguards products during transportation and handling. EPS foam offers a stable, long-term thermal insulation performance despite initial fluctuations caused by the manufacturing process [4]. In Fig. 1, the primary characteristics of EPS have been shown.

EPS's versatility extends across various applications, encompassing packaging, construction, and disposable products. Norway pioneered EPS as a lightweight material for road construction on weak subgrades in the 1960s. Recognizing its additional benefits, EPS has gained popularity as a substitute for traditional improvement methods in various civil engineering projects. Its ability to reduce vertical and lateral stresses while maintaining adequate compression strength has led to a surge in its application in embankment construction, slope stabilization, retaining structures, bridge abutments, and underground infrastructure [5]. In construction, it is used as insulation for walls, roofs, and floors, as well as ornamental features such as crown molding and baseboard [6]. It also serves as the framework for constructing lightweight filling structures and geofoams [7]. Apart from construction, EPS is employed in manufacturing disposables, floats, automobile parts, farming, craft, and model making. As EPS is lighter, they are ideal for packaging goods such as electronics, appliances, furniture, food items, machinery, equipment, and other industrial

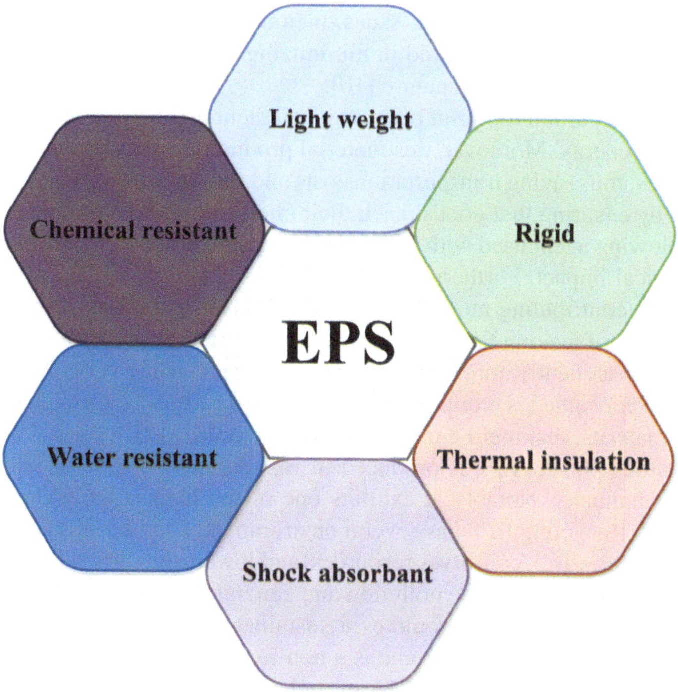

Fig. 1 Properties of EPS

products. EPS's versatility has led to its widespread adoption across diverse industries. Nevertheless, its disposal presents environmental concerns, necessitating efforts to promote sustainable recycling and alternative materials.

1.2 Importance of Recycling EPS

Waste recycling has become an increasingly pressing issue for many nations worldwide, particularly in developing countries. Expanded polystyrene (EPS) is commonly used as a shock absorber in electrical and electronic appliances. Therefore, with the increased production of electrical appliances EPS wastes increases significantly [8]. Despite attempts to increase recycling rates, EPS wastes regularly go to landfills. Improving recycling rates has been made, but still very low, considering that recycling could attain a whopping level of 100% with good formulation. EPS litter harms our planet because it spills into rivers and long degrading periods in landfills [9]. By poisoning the environment, it can lead to the death of animals. Research says that micro polystyrene beads can harm flies, among other organisms, and promote the accumulation of toxic substances. Proper waste management techniques should be adopted to resolve these issues, including recycling EPSs and disposing of them appropriately. This will aid in minimizing landfill waste, thereby reducing the negative impacts of EPS on nature [10].

EPS has a low carbon footprint because its efficient production method consumes less water and energy. Moreover, this material produces less waste due to its protective properties, thus saving transportation costs and resources. It also aids in reducing food waste by ensuring that goods reach their clients in the best possible conditions [11]. The blowing agent used with EPS is pentane, a natural hydrocarbon with negligible ecological impact. Furthermore, since it is lightweight, EPS needs less transportation fuel, contributing more to its small carbon footprint. The environmentally friendly process of making EPS involves steam as its main ingredient and uses water repeatedly. Consequently, minimal wastage is generated while very little of the total amount of oil consumed is required. EPS has a lower carbon footprint than any other packaging material, making it a more sustainable choice [12].

A ubiquitous plastic, EPS is produced in vast quantities globally, exceeding 14 million tons annually. Notably, it exhibits one of the highest recycling rates after industrial use. Recycling EPS has several environmental and economic advantages. It reduces landfill waste, conserves resources, and lowers the carbon footprint. Additionally, it helps prevent marine pollution and generates revenue while creating jobs [13]. Overall, recycling EPS promotes a sustainable and circular economy. As a petroleum-based product, polystyrene is a non-renewable resource. Recycling EPS mitigates long-term environmental harm and presents financial opportunities for companies [14].

2 Understanding Mechanical Recycling

2.1 Definition and Principles

Mechanical recycling is an efficient way to manage waste. In mechanical recycling, waste materials are turned into new products through physical means like collection, sorting, washing, drying, shredding, re-granulating and compounding without significantly altering their chemical composition [15]. Methods such as extrusion, injection molding, or compression are employed in mechanical recycling. Put another way, and it requires carefully separating and washing out waste streams to preserve the quality of recycled materials for a more sustainable and resource-efficient future. Figure 2 shows the mechanical recycling process.

Mechanical recycling can be classified into pulverization, compaction, and extrusion. The purpose of pulverization is to reduce polystyrene into smaller pieces for remolding or as a filler. Thereby reducing waste, conserving resources and benefiting the environment. To compress EPS, hydraulic and screw compactors are commonly used [16]. Despite both means being able to effectively decrease EPS's size, screw

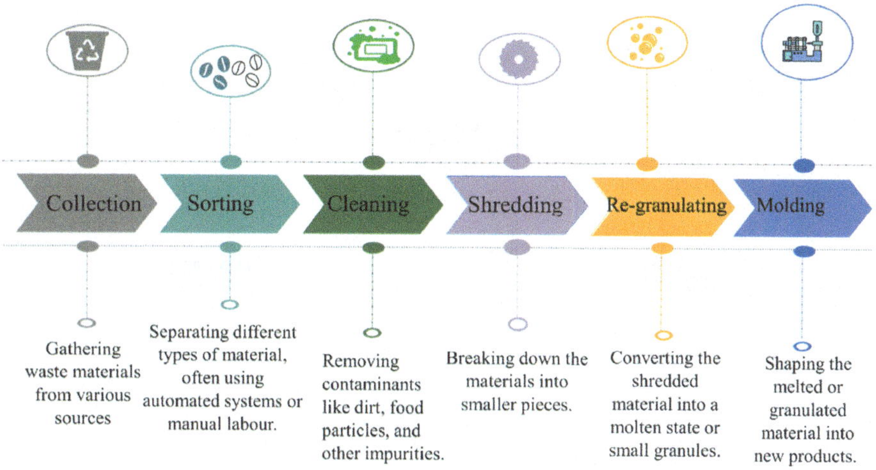

Fig. 2 Mechanical recycling methodology

compactors are usually more effective because they work better under pressure when compressed. Due to this precise pressure control, material is compressed more uniformly and effectively. Expanded polystyrene (EPS) is melted down and shaped into new goods like sheets or pellets using extrusion and injection molding techniques. There are three main types of polystyrene: expanding foam, solid and reinforced. However, solid polystyrene can be easily recycled mechanically because of its relatively dense structure. In contrast, expanded polystyrene must go through a particular size reduction process before recycling owing to its porous, cellular nature [17].

Comparison with Other Recycling Methods (Chemical, Thermal, Biological)

Recycling is a way to conserve resources, reduce landfill waste, and minimize environmental impact. They can be categorized into various types. As shown in Fig. 3, all four types can be further categorized into many various sub-sections.

To summarize, in mechanical recycling of EPS, unwanted EPS is reduced in volume, melted, and then molded into useable goods using machinery (primarily extruders and compacting machines). Chemical recycling, on the other hand, is comprised of employing the appropriate solvents to reduce EPS waste in its monomer or solution. However, thermal recycling involves heating EPS waste to a controlled high temperature to break down long-chain hydrocarbons [8]. Finally, the waste materials are decomposed and transformed into useful products in biological recycling. Although mechanical recycling is the more popular process of recycling, the elevated proportion of intricate composite materials, coupled with stringent quality standards for secondary raw materials, constitute the primary obstacles hindering the attainment of a higher recycling rate through mechanical processes [18]. Chemical

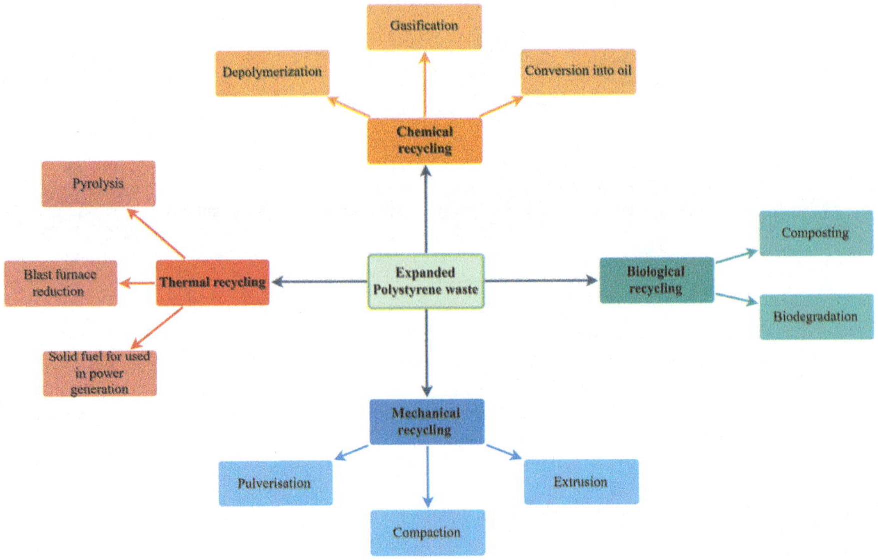

Fig. 3 Types of recycling process

recycling can process materials with contaminants or low quality that are difficult to recycle by mechanical means. It can also produce high quality materials. However, they often require more energy than mechanical recycling and can also produce harmful greenhouse gas in the production process. Thermal recycling can process various waste materials, including those unsuitable for mechanical recycling, and it can produce energy. However, their environmental concerns, high capital costs and limited waste suitability makes them difficult for wider applications. Biological recycling is the most sustainable and natural recycling process among the four [19]. This method also helps in improving soil quality. Nevertheless, it is a slow process. Also, the constraints of limited material suitability, quality limitations, and infrastructure requirements hinder the widespread adoption of biological recycling. In Fig. 4, a comparative study of recycling methods has been depicted. Mechanical recycling is a method that transforms waste into useful products via its physical change.

2.2 *Advantages and Disadvantages*

Mechanical recycling has its own advantages and limitations. Mechanical EPS waste recycling is more environmentally friendly than chemical or thermal methods, as it does not emit volatile organic compounds (VOC) or offensive odors like petroleum-based EPS solvents. Mechanical recycling is more energy-efficient than thermal

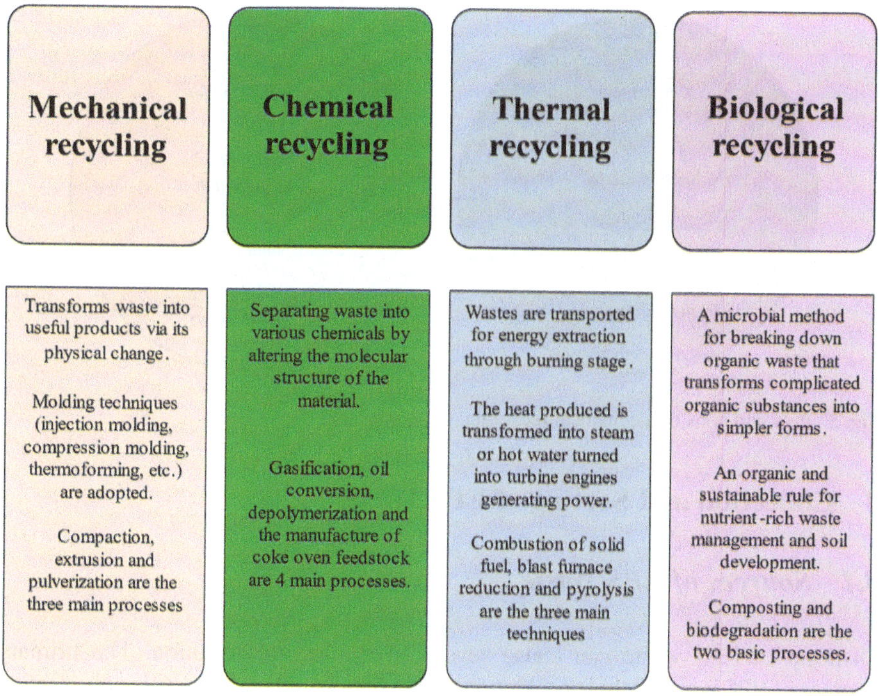

Fig. 4 Comparison among recycling methods (mechanical, chemical, thermal and biological)

recycling because it doesn't degrade EPS. The mechanical method of recycling EPS trash is cost-effective due to its direct approach. This recycling technology produces a variety of items, including plastic lumber and EPS insulating products. This approach is more convenient to recycle because of its available infrastructure and well-known and straightforward methodology. It is also less time-consuming than biological recycling. On the other hand, the cost and quality of the finished product are the main drawbacks of recycling EPS mechanically [20]. It is important to remember that recycled materials frequently have worse quality than virgin materials, and mechanical recycling of EPS is not always the most economical option. Contamination is a significant challenge in mechanical recycling, as it can reduce the quality of recycled materials and make them unsuitable for certain applications. This can also lead to equipment damage. In Fig. 5, the pros and cons of mechanical recycling has been shown.

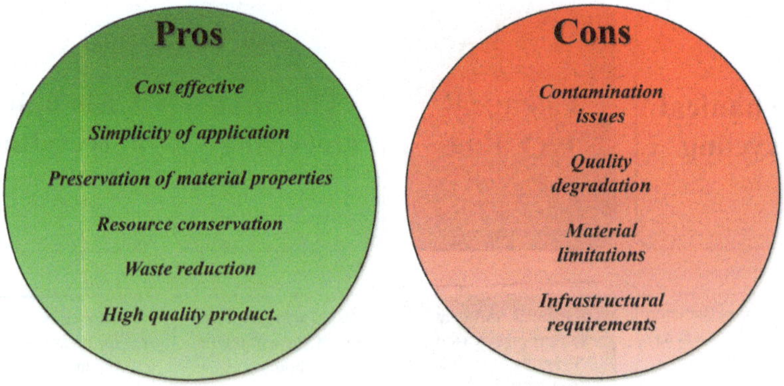

Fig. 5 Pros and cons of mechanical recycling

3 Collection and Sorting of EPS Waste

3.1 Sources of EPS Waste

Disposal of EPS waste can cause severe environmental pollution. The primary sources of EPS waste include packaging, construction and building materials, disposable products, floating devices, automotive components, etc. All these sources can be divided into two categories: industrial sources and consumer sources.

Industrial Sources

Industrial sectors are significant contributors to EPS waste due to their reliance on packaging and insulation materials. Industries like electronics, automotive, appliance, furniture, construction, food and beverage, and chemical/pharmaceutical sectors produce a lot of EPS waste primarily because they use it for packaging and insulation. Landfills are overwhelmed by the large amounts of industrial EPS waste. This waste can be harmful to the environment. To reduce the impact, these industries should try to use less EPS, recycle more, and look for alternative materials that don't harm the environment.

Consumer Sources

Consumers are a significant source of EPS waste due to the widespread use of EPS products in everyday life. Common consumer sources include food packaging, furniture packaging, electronics packaging, refrigerator, stove and appliance packaging, etc. Many disposable products, such as cups, plates, and cutlery made of EPS, are also used in our day-to-day lives. EPS is used in decorative items like crown molding and baseboards. These items are often discarded after single use, contributing to the accumulation of EPS waste in landfills and the environment. To minimize the impact, consumers can reduce their use of EPS products, recycle whenever possible, and support the development of sustainable alternatives.

3.2 Collection Methods

Effective collection methods are crucial for managing EPS waste and minimizing its environmental impact. Collection of waste is a labor-intensive process. It is also a very costly process. About one third of the total waste management cost is allocated for waste collection [21]. There are different methods of waste collection. Among them, drop-off centers and curbside collection are discussed below.

Drop-Off Centers

Specialized centers for waste recycling are a convenient location to throw away EPS from both home and business organizations. These centers have specialized equipment and methods for proper EPS waste handling. For instance, they may utilize processes such as pelletizing, shredding or densification to prepare EPS for recycling. The availability of these recycling hubs presents a simple and affordable option for all types of waste creators.

Curbside Collection

In this specific method of waste collection, EPS waste is separated from other recyclable materials or wastes. Municipal administrations provide special containers for EPS rubbish, and collection usually occurs on certain days or at specific times. Curbside collection is an efficient and convenient form of waste disposal that supports community recycling programs.

There are several challenges in EPS waste collection. The significant barriers facing EPS waste collection include contamination with other materials, shortage of manpower to collect large amounts of refuse, and lack of infrastructure. To overcome these challenges, investment should be made in EPS recycling infrastructure, efficient collection systems should be established, and proper disposal should be promoted.

3.3 Sorting Techniques

Sorting expanded polystyrene (EPS) waste is a significant step in the recycling process. It ensures that the material is clean and free from contaminants, making it suitable for reuse in manufacturing processes. Sorting techniques are categorized into the following methods.

Manual Sorting

The manual sorting of EPS waste is a visual activity. Each piece of EPS waste must be inspected visually to unearth any pollutants or impurities. Included in this are also plastics, paper and leftover foods. This method requires sorting the garbage items based on their size, shape and appearance. In addition, they can be classified using colors that help distinguish them into different kinds or classes.

Automated Sorting Technologies

Automated sorting systems use advanced technologies to sort items quickly and accurately based on their chemical or physical characteristics. Such methods are useful for very large waste streams (for example, in expanded polystyrene recycling). The referred technology employs various important approaches such as optical sorting, density separation, X-ray sorting, and magnetic sorting. In optical sorting, sensors like infrared radiation identify variations in the colors, shapes and surface properties of materials thereby enabling proper separation. Flotation or air classification techniques can be used to separate EPS because it has lower density than many other substances. Magnetic separators assist in recovering metal contaminants from EPS garbage. X-ray technology can be applied to identifying substances depending on their atomic arrangement. This is particularly helpful for classifying materials with similar looks, but density or composition differs. By making recycling processes of EPS more automated, waste management programs can be made generally more sustainable and effective [22].

4 Pre-processing of EPS Waste

The expanded polystyrene recycling process needs preprocessing which is one of its most significant aspects since it assists in making the waste reusable and reproducible. It consists of various processes aimed at eliminating contaminants, lowering mass of material and enhancing its standard. The sequential description of the pre-processing of EPS waste has been discussed here.

4.1 Cleaning and Contaminant Removal

EPS is cleaned using various methods to remove contaminants such as oil, dirt or other substances. Moreover, EPS waste also undergoes chemical treatment with some types of impurities removal. In this case, detergents and solvents are used. One more important step after washing and drying should be taken to EPS; otherwise, its processing and quality will be affected by moisture content.

Techniques for Cleaning EPS

Cleaning up expanded polystyrene waste for recycling is a vital step that helps to eliminate impurities which may interfere with recycling and affect the quality of the reclaimed item. Some cleaning techniques are discussed below.

i. **Physical cleaning:** During physical cleaning processes, expanded polystyrene surfaces are cleaned up first with water and detergents to eliminate surface impurities. Afterwards, if some of these impurities persist, they are manually or mechanically scrubbed off. Subsequently, dirt and loose particles are removed using compressed air.

ii. **Chemical cleaning:** The three different techniques that compose chemical cleaning include alkaline, acidic, and solvent cleaning. Solvents are utilized to dissolve and eliminate specific contaminants like adhesive substances or oils. Alkaline solutions are used to remove grease and acidic contaminants, whereas alkaline deposits or pollutants are removed through acid cleaning.

iii. **Advanced cleaning techniques:** The cleaning technology has greatly changed by the advanced methods like Plasma and ultrasonic cleaning. The ultrasonic cleaning uses high frequency sound waves to enable getting rid of impurities from the expanded polystyrene surfaces. Alternatively, the use of ionized gas in Plasma cleaning enhances the surface characteristics of expanded polystyrene thus eliminating impurities.

Handling Contaminated EPS

Contaminated EPS contains foreign materials which makes recycling difficult. It's crucial to carefully identify and separate tainted EPS [23]. The contaminated EPS must be stored safely to prevent additional contamination or any accidental release. For efficient transportation and storage, the weight of the impure EPS should also be as low as possible. The right cleaning methods must be applied for the removal of contaminants from the waste. Sometimes, disposal at a landfill may be the only feasible option when recycling and complete cleaning are not possible. The last resort for contaminated EPS is incineration, where recycling or safe landfill options do not exist. Alternatives may take innovative forms, such as chemical recycling and energy recovery. By chemical recycling, the polluted EPS is broken down into its base molecular units so that it can be reused in other products. Energy recovery involves turning contaminated EPS into a fuel through processes such as gasification or pyrolysis [24]. Contaminated Expanded Polystyrene (EPS) can be managed properly by using these methods and adhering to safety protocols. This will reduce landfill dumping and promote a more sustainable approach to the management of waste.

4.2 Size Reduction

EPS are shredded into smaller sizes to ensure easier handling of waste and storage. It is mainly performed by two methods: shredding and grinding.

Shredding

Shredding is used to reduce materials into smaller pieces. The process is mostly used for making waste disposal easier, as well as improving transportability and recyclability of different waste materials. Some advantages of shredding include significant size reduction, better recycling processing and prevention against accidental release of toxic substances.

Grinding

Grinding is a fundamental process in many industries, and it involves the removal of material from a workpiece by means of abrasive tools. It is crucial to recycle EPS as it reduces the materials' volume and prepares it for re-use in various applications. In this process, EPS waste, employing grinding mills; specialized machines with rotating knives or hammers are crushed into smaller pieces.

There are several challenges in pre-processing. Cleaning and sorting EPS waste can be complicated processes mostly due to contamination with other materials. Preprocessing becomes more challenging due to various densities, colors and shapes of EPS files used. Manual sorting and cleaning could also consume a lot of human resources and turn out ineffective for large quantities of refuse. However, preprocessing has witnessed major advancements which have overcome several barriers [25, 26]. Modern sorting improves both speed and precision through state of the art sensors as well as algorithms that enhance accuracy levels when dealing with different wastes streams. Additionally, greener chemical treatment methods for destroying contaminants have been developed to make sorting less toxic in general. A further step towards better waste pre-processing is represented by energy-efficient processes designed for drying EPS refuse. This way effective grade EPS wastes can be obtained from disposable EPS enabled products thereby minimizing their adverse effects on society while supporting circular economy model.

5 Mechanical Recycling Process

Mechanical recycling is the process whereby waste materials are recycled for physical operations. For plastic to be recycled, it must first be shredded or ground into granules, impurities must be removed, and flakes must be separated. The melted recycled plastic is extruded into uniform pellets that can then be used in various production techniques such as extrusion, injection molding and blow molding. The properties of plastics depend largely on their molecular weights; usually, high molecular weight plastics show high glass transition and melting temperatures, and they are tougher, more elastic, stiffer, more resilient, and viscoelastic. The processing time and number of recycling cycles can also influence the molecular weight of the recycled plastic. Repeated exposure to high temperatures, prolonged periods, and intense shearing can degrade polymeric materials, sometimes leading to unintended chemical reactions that either increase or decrease the molecular weight of the final product [27]. These structural changes affect both the material's flow and mechanical properties. While manufacturers may add additives to restore material properties or ensure homogeneity, this often increases costs and can hinder future recycling efforts. In Fig. 6, the mechanism of mechanical recycling is shown.

Fig. 6 Mechanism of mechanical recycling

5.1 Densification

Compaction Methods

Densification is a process that enhances the density of materials which can be achieved through various methods. These include compression, which reduces volume through applied pressure; heating, which induces particle vibration and subsequent compaction; and incorporating denser substances to elevate overall density. Compaction methods are techniques used to reduce the volume of materials, often by increasing their density. There are numerous compaction methods. The classification of compaction methods is shown in Fig. 7. Mechanical compaction involves using rollers, rammers, or vibrating plates to compress materials like soil or asphalt. Thermal compaction methods include incineration and pyrolysis, which burn or heat

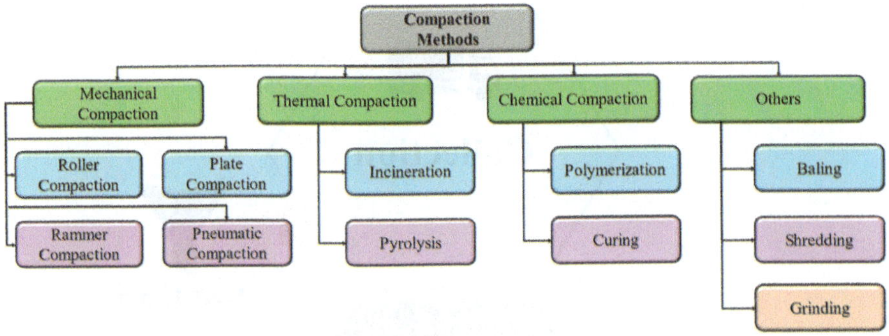

Fig. 7 Classification of compaction methods

waste materials to reduce their volume and generate energy. Chemical compaction uses chemical reactions to cross-link or polymerize materials, reducing their volume. Other methods include baling, shredding, and grinding to make materials more compact.

The choice of compaction method depends on the type of material, the desired level of compaction, and the specific application. Densification is employed across diverse industries, including manufacturing, construction, and waste management, to optimize material properties, mitigate costs, and augment efficiency.

Densification Equipment

Different equipment is used for different compaction methods. Roller compaction uses heavy rollers to compress materials like soil or asphalt. Rammer compaction uses a rammer to compact materials in confined spaces, such as trenches or foundations. A vibrating plate is used in plate compaction, and in pneumatic compaction, air pressure is used to compact materials like soil or sand. Thermal compaction relies on specialized equipment to burn waste and harness the resulting heat. Incinerators, pyrolysis, gasification, and heat recovery systems are commonly used [28]. The choice of equipment depends on the type of waste and the desired outcomes. Chemical reactors, mixers, temperature control systems, pressure vessels, and separation devices are commonly employed in chemical compaction. The specific apparatus utilized in chemical compaction is contingent upon the nature of the materials being processed and the desired outcomes. Besides, ballers, shredders, mold and many other equipment are used in densification.

5.2 Pelletizing

Process Description

The method by which recycled materials are organized systematically onto pallets for improved storage, transportation and processing is called palletization. Palletization optimizes operational efficiency since it reduces manual handling and increases space usage in accordance with its fullest potential. Furthermore, it aids in transport since it permits the use of varied means of transport without any hitch whatsoever. Besides, pallets protect recycled materials against damage during transit and storage periods. By utilizing this technique, recycling plants can enhance their efficiency levels considerably, cut down on operational costs, andromote the success of mechanical recycling ventures [29].

Equipment Used for Pelletizing

The equipment employing palletizing in mechanical recycling includes forklifts for lifting and moving pallets, pallet jacks for manual or powered transportation, pallet wrappers for securing materials with plastic film, and pallet dispensers that automatically distribute empty pallets. These facilitate efficiency in handling, storage and transportation of recycled materials. Equipment used in palletizing is shown in Fig. 8.

5.3 Extrusion

Extrusion Techniques for EPS

The gathered EPS waste materials can be shaped by making use of the extrusion method. It is an adaptable technique for manufacturing a variety of items with dependable quality and exactness. Various approaches are included in extrusion techniques that are used to shape EPS into versatile products. The most prevalent technique, single-screw extrusion, utilizes a solitary rotating screw to propel EPS material through a die. Twin-screw extrusion, employing two rotating screws, offers enhanced mixing and process control. Co-extrusion, a method involving the simultaneous extrusion of multiple materials, facilitates the creation of layered or composite products [30]. Sheet extrusion produces flat EPS sheets, commonly utilized for insulation and packaging. Profile extrusion yields shaped EPS profiles like pipes, rods, or channels. Foam extrusion, a technique that generates expanded EPS foam with a closed-cell structure, is frequently employed for insulation and packaging applications. The selection of an extrusion technique is contingent upon the desired product attributes, including density, shape, and size. Figure 9 shows the general extrusion mechanism of plastics.

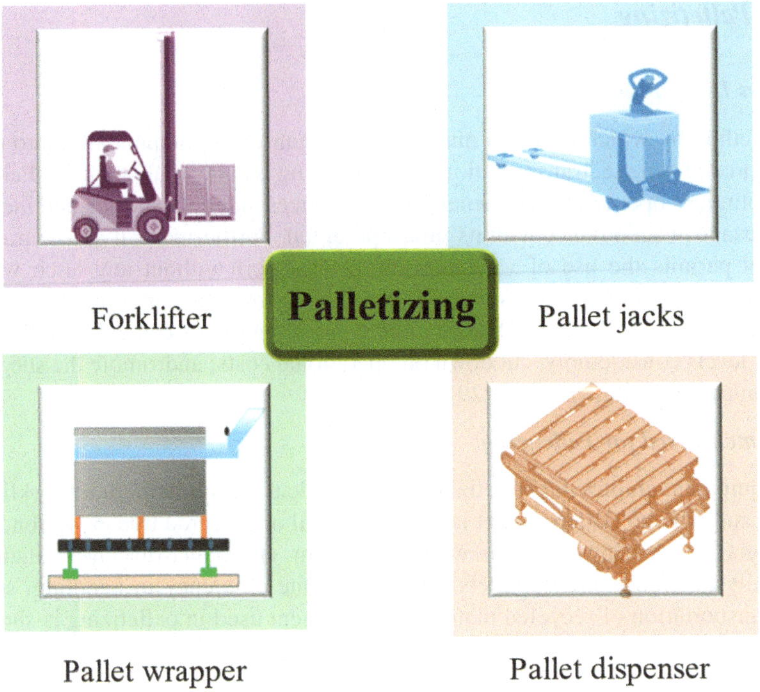

Fig. 8 Equipment used for pelletizing

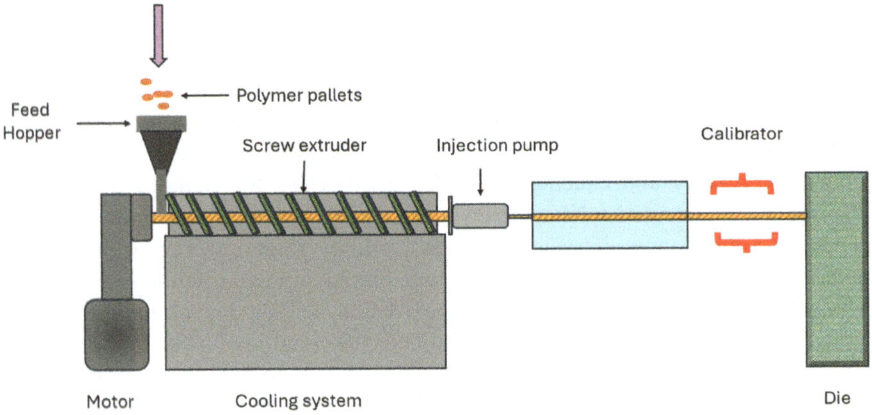

Fig. 9 Extrusion mechanism

Parameters and Optimization

Extrusion techniques are essential for molding EPS into a variety of items. The regulation of different constituent parts ought to be done with utmost precision to obtain the best results. The choice of screw design, temperature, pressure, die design, and speed of feeding together with vacuum application and cooling methods are vital components [31]. It is possible to produce high–quality EPS products through minimizing energy usage as well as waste in this way: In detail experimentation, simulation, statistical analysis and process control ought to be carried out against these parameters which will culminate into the characterization of materials.

6 Properties of Recycled EPS Products

The potential applications of rEPS in various fields, such as construction materials, polymer composites, and insulation, requires an understanding of its properties and associated limitations. By addressing these challenges, researchers and industries can capitalize on the benefits of rEPS and develop more sustainable and innovative products.

The mechanical properties of rEPS might even be better than those of pristine EPS. A study investigated the use of recycled EPS as a partial replacement for sand in rendering mortars. While mortars with added EPS have improved fluidity, their mechanical properties were negatively impacted. Recycled EPS flakes showed better performance than virgin EPS beads in terms of compressive strength (17.5 versus 13.5 MPa) and flexural strength (5.5 versus 4.5 MPa). Water absorption rates varied greatly, whereas adhesive strength remained the same as the reference mortar's. Consequently, these conclusions point to the possibility of using recycled EPS instead of sand for rendering mortars since it provides better fluidity and similar mechanical performance [32].

Eco-friendly polymer composites can be fabricated by using rEPS as the matrix which can contribute to lessen the environmental impact in the present context. Abdel-Hakim et al. [33] produced sawdust filled rEPS composites to assess the sound absorption, mechanical properties and biodegradability. The composites showed good sound absoprtion behaviour proportional to increasing sawdust content, however it had a negative effect on the mechanical properties. 20% inclusion of sawdust produced the best mechanical characteristics with 16.7% increase in tensile strength and 14.1% in flexural strength. At higher loadings beyond 20%, aggregation of sawdust particles initiated causing deterioration in performance. This was validated using scanning electron microscopy (SEM). Biodegradability of the composites were performed through soil burial for 90 days. Interestingly, the weight loss increased with the increase of filler loading.

The properties of cement mortars partially substituted with recycling expanded polystyrene (EPS) was assessed by Petrella et al. [36]. The intention was to develop lightweight thermo-insulating composites having minimal water absorption and high hydrophobicity through an environmentally friendly and economical approach. The cement mortars were either partially or entirely replaced with rEPS of various sizes. While rEPS mortars had a higher thermal insulation due to their reduced density, there was also an associated loss in mechanical strength. Nonetheless, sand-EPS mixtures with 50% EPS were able to achieve a trade-off between thermal insulating capacity and the required mechanical resistance. In addition to being hydrophobic and having reduced water penetration rates, EPS mortars showed improved performance particularly for 2–4 and 4–6 mm size rEPS pieces. This was due to the low surface energy of rEPS alongside good particle distribution. These composites have lower densities, behave as insulators against heat transfer, making them more applicable indoors without harming nature and as a result, they become environmentally well oriented. Recycled EPS can also be effectively utilized in conjunction with epoxy matrix composites. A study observed how utilizing expanded polystyrene (rEPS) bubbles coated on their surfaces would influence the characteristics of epoxy matrix materials (Fig. 10a). Although rEPS reduced the density and mass, it also decreased the impact strength. The surface gloss was modified while hardness was almost constant. This study highlights that achieving a uniform distribution for rEPS is crucial and optimizing the amount for use in epoxy resin composites. Therefore, rEPS can be used to enhance lightness in epoxy resin composites, but care must be taken not to compromise on targeted mechanical properties [34].

Insulation materials made from rEPS can be a viable alternative for low-wage earners. A study proposed a new insulating material that was produced from recycled EPS, cement, plastic additives, and water (Fig. 10b). Characterization tests also showed that the obtained thermal conductivity range of 0.0603–0.0706 W/m K is competitive with commercial materials, which means it has the potential to be used for insulation. It also passed the electrical overheating test up to 960 °C and water absorption rate test. The material's mechanical characteristics, such as bending load at break (18.3 N) and compressive strength (126.4 kPa), are suitable for non-structural applications. These results indicate that the proposed material is feasible and cost-efficient in enhancing thermal comfort and decreasing energy use in low-income homes [35].

7 Applications of Recycled EPS

Presently, rEPS find application in packaging mainly because of its good shock-absorbing nature and heat insulating property. For instance, foam peanuts produced from rEPS are always used to pack and fill the empty spaces of shipping containers so that delicate goods are well protected from any sharp forces as they are shipped. Also, the rEPS sheets can be easily cut and molded to fit the product that is being packaged to suit the shape and size of that product. In the automobile sector, rEPS are applied to manufacture lightweight and high identifier parts like bumpers and dashboards [37]. These components assist in decreasing the overall weight of the vehicle thus enhancing the fuel consumption as well as maneuverability. In addition,

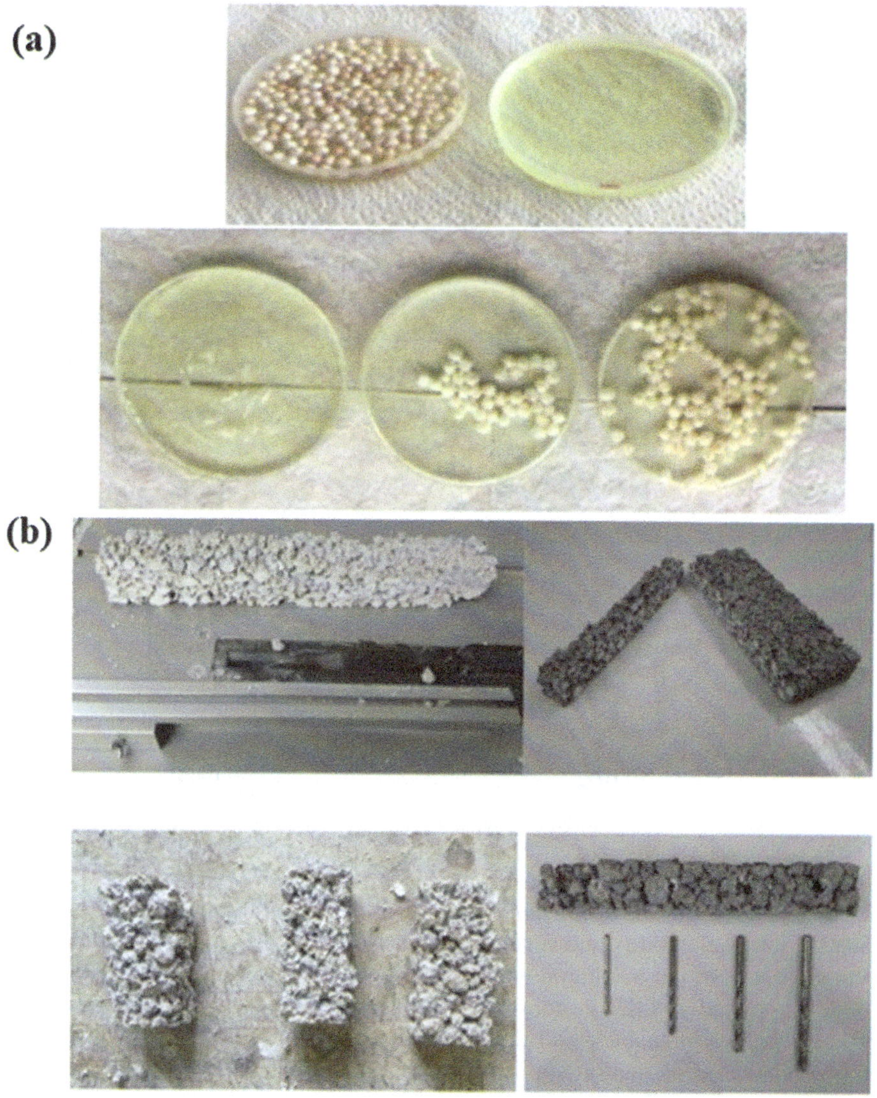

Fig. 10 **a** rEPS infused epoxy/PA6 composites [34] and **b** thermal insulation material fabricated using rEPS [35]

rEPS are applied in seat cushions, making drivers and passengers comfortable while on a drive. rEPS are used in medical applications mainly because they are non-toxic, biocompatible, and lightweight. For example, when developing a comfort, easy-to-breath through oxygen mask, rEPS is used. Also, rEPS are applied to lab apparatus and convectional samples to protect them against damage when they are transported or stored [38]. Figure 11 illustrates some common uses of rEPS.

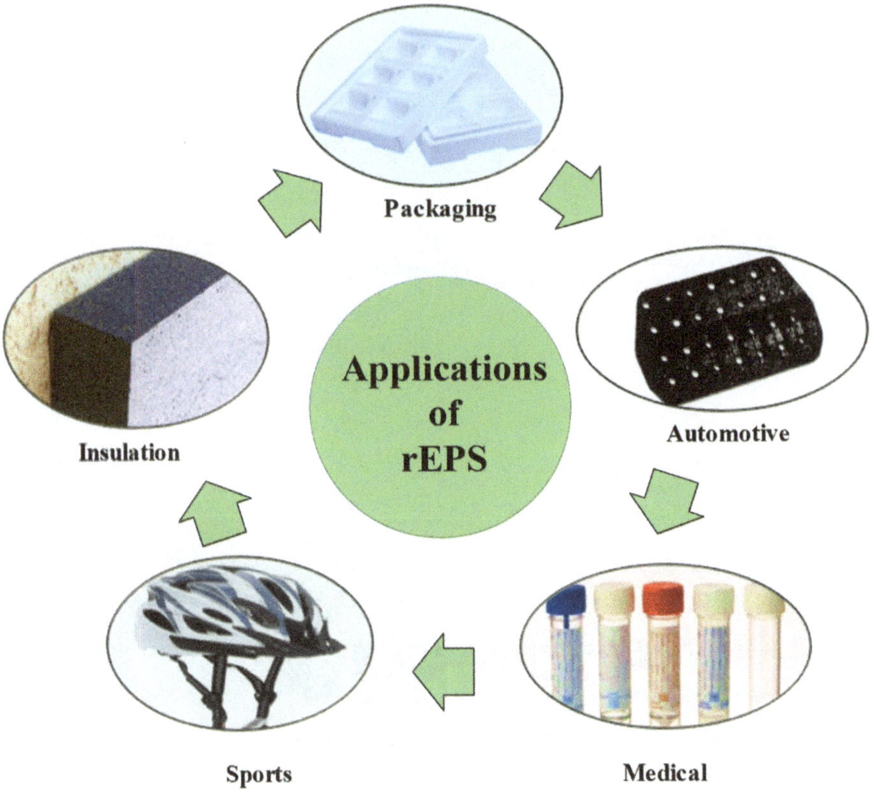

Fig. 11 Generic applications of recycled expanded polystyrene (rEPS)

rEPS are one of the most consumable products in the construction industries due to their flexibility, durability and versatility in its use. Other uses of rEPS are discussed below apart from the insulation boards and concrete aggregate uses. For instance, it can be applied as filler in constructions of roads as it has the added advantage of being lighter to enable good drainage. Moreover, the rEPS are also employed in geotechnical construction applications, such as soil stabilization and control of erosion. Its other attributes and general characteristics include very low weight and high-water absorption capacity, making it the right material in such uses. In addition, rEPS are used to protect pipelines and electrical cables by offering them insulation and shielding against harm in underground places [41]. They can also be used as subflooring, meaning they create a flat surface on which finished forms of flooring can be laid. Consequently, rEPS encompass numerous advantages in construction through energy efficiency, sustainability, and increased building performance [8]. Figure 12 shows the use of rEPS in construction materials.

When used in the asphalt mixture, the recycled EPS can improve the waterproofing capability for better resistance against moisture and thus improve the pavement's

Fig. 12 Usage of rEPS in construction materials: **a** lightweight concrete sandwich panels [31], **b** rEPS infused self-consolidating lightweight concrete (SCLC) [39], **c** rEPS light weight block [40], **d** light weight earth block [28]

service performance. Also, EPS has the added effects on the rheological characteristics of the asphalt making it have higher resistance to the rutting and cracking [42]. In the paint industry, EPS could be used as a binder material in emulsion paints, which should be environmentally friendly in contrast to conventional binders. This can help reduce the amount of harm done to the environment in the production of paints while, in the same process the quality of paints may be enhanced [43]. Moreover, when the most important characteristics must be enhanced, EPS can be mixed with natural rubber and produce new products. These blends can be applied in automotive parts, footwear, and other consumer products with enhanced durability, flexibility, and other properties. In general, the recycling and reuse of EPS waste will not only help preserve the world from being filled and become a dumping ground for wastes but also offer a significant value in energy efficiency and optimum material performance [44].

8 Challenges and Solutions

8.1 Common Hurdles

Although mechanical recycling of EPS offers certain benefits, several challenges are still present which hinders it's widespread adoption across the recycling industry (Fig. 13). Therefore, unique strategies are required to address these limitations to establish a more reliable process for recycling of EPS. One of the primary concerns is the lower quality of recycled EPS compared to the unprocessed material. This restricts its application in various fields and reduces its selling price. To mitigate this issue, advanced technologies are required to retain different properties such as density, strength, and thermal conductivity [45]. Research groups and institutions can take joint initiatives to develop more sophisticated equipment to ensure the excellent condition of recycled EPS. Another significant challenge is the higher operational costs associated with mechanical recycling compared to other techniques. The high costs can be lowered if efficient processes are implemented. This includes but is not limited to energy optimization and waste minimization. Higher end power-saving machines can be utilized effectively to control these costs. Additionally, government incentives and subsidies can attract investors in mechanical recycling facilities and make them more economically viable [46]. Due to the high upfront costs, scaling up mechanical recycling facilities is tedious. Partnerships between governments, industries, and NGOs can improve the development of a robust recycling infrastructure. Another significant issue is the high level of contamination associated with mechanically recycled EPS. The EPS collects a lot of dust and debris, some of which are deeply embedded within the beads. This is comparatively easier to remove with other techniques. Strict quality control and advanced sorting technologies can help mitigate this issue. The disposal of EPS at the right place can be encouraged at locality levels to avoid unnecessary contamination as most used foams are collected from waste where all types of garbage are present [15].

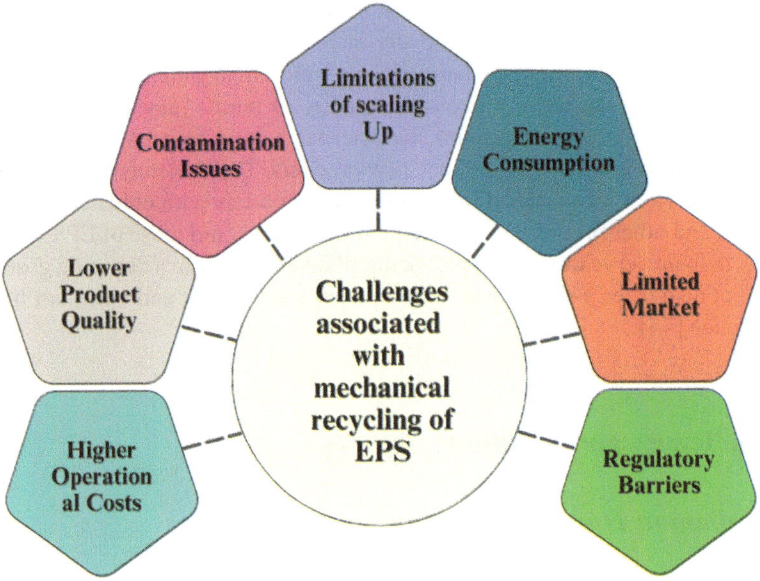

Fig. 13 Challenges of implementing mechanical recycling of EPS

Finally, limited market demand for recycled EPS can affect its commercial viability as customers may have dilemmas about the overall quality of the product. Enhanced marketing, targeted marketing efforts and product development can help increase consumer awareness and acceptance of recycled EPS products. The relaxation and amendments of government policies will allow the integration of recycled EPS into sectors which currently do not hold authorization for its use [47]. Through the address of the challenges with a combination of technological improvements, government support and collaboration of different industries the mechanical recycling of EPS can be more suitable for optimal management of EPS waste. This will contribute to a cleaner environment and promote economic opportunities [48].

8.2 Technological Advances

EPS recycling has seen significant innovations in recent years, especially in the areas of sorting and processing. These advanced techniques have helped the recycling process to be more efficient, cost-friendly and sustainable. Optical sorting is one of the most effective technologies applicable in recycling expanded polystyrene (EPS). Through application of high-speed camera and special sensors, the EPS can be segregated from the balance waste flow through this process. On the conveyor belt, the EPS wastes are subjected to various tests that include density and surface texture, among others. Thus, the measurements performed in the system allow one to aim the EPS for further utilization and discharge it. This technology not only improves the process of recycling EPS but also reduces the probability of contamination of other forms of recycled items and, therefore, has positively impacted waste management industry [49].

Near-Infrared (NIR) Spectroscopy is another technique which is quite helpful in EPS recycling. Near-infrared identification and recognition technology identifies and recognizes the absorption and reflection of near-infrared rays in the waste materials. This type of spectroscopy will determine the chemical makeup of a material to separate EPS from other plastics and contaminants. It is then possible to distinguish the EPS from the rest of the waste material by means of the spectral signature that the NIR sensors have for sorting and recovering the material. NIR offers a non-destructive and rapid method for EPS waste analysis, making it a valuable addition to optical sorting systems in EPS recycling facilities. AI algorithms can be used to correctly sort and classify data obtained from optical sorting systems as well as NIR spectroscopy of EPS, whereas robotic systems can be used to perform variety of functions increasing productivity and decreasing the number of employees needed. When integrated together with industrial EPS recycling techniques, such facilities can substantially improve the performance of EPS recycling to support the creation of a circular economy and reduce pollution [50].

9 Case Studies and Success Stories

9.1 *Municipal and Corporate Initiatives*

California

A local organization in Tuolumne County, California known as Master gardeners pick EPS waste that the citizens have disposed of. The recycler only takes clean food containers which must be placed in clear plastic bags. Tightly packed grade polystyrene which is used commonly to pack electronic items, and other products may be sorted and packed in colored plastic bags. They suggest that one should donate only clean packaging foam and rigid, white, foam insulation and ensure that all the items do not contain tape, labels, stickers, paint, cardboard and other things that are not polystyrene. They also state that food containers must be clean and free from any form of food residues as well as grease. Among them, there are restrictions on the acceptance of bubble wrap [51].

Texas

Texas-based Dart Container has launched a new EPS foam recycling center in Ellis County, Texas. The 24/7 drop-off center accepts foam food service containers with the chasing arrows symbol of number six. After that, the foam is compressed and taken further to another company where the foam is used to manufacture other products. This organization seeks to prevent foam from being dumped to the landfills and encourage foam recycling for such uses as picture frames and moldings. They emphasize the environmental benefits of foam recycling, including reduced waste in production, landfill stability, and clean burning in energy-from-waste facilities [52].

Home Depot

Home Depot is very active in environmental conservation, embodied in its recycling program, which includes EPS (Styrofoam). Their program is about collecting EPS from appliances packaging, palletizing or compacting into blocks and sale to producers of insulation among other products. This effort is to decrease waste and promote a circular economy in society. By recycling, The Home Depot is therefore playing a worthy role in ensuring that EPS does not end up in the landfills and encouraging its reuse [53].

9.2 *Lessons Learned and Best Practices*

Any successful EPS recycling scheme will require massive cooperation between the municipalities, various businesses, and the recycling companies. The public should be educated on the necessary and correct ways of disposing of the waste. Some of the key factors include curbside collection as well as the availability of drop-off centers that act as focal points in EPS collection [54]. To keep the program going, markets for recycled EPS must be developed by buying products from companies that use post-consumer material and/or coming up with new products. Monitoring and data collection are important to assess a program's strengths and weaknesses and to verify its efficacy. These guidelines help communities achieve successful EPS recycling programs that positively impact waste management and the environment. Furthermore, increasing public awareness through health promotion activities and involving the public through EPS recycling campaigns and other community activities would also strengthen the EPS recycling programs. When people are motivated to embrace the physical act of participating in the recycling processes of their neighborhoods, then this will be very effective in transforming the face of the environment [55].

10 Future Directions and Opportunities

Presently, the market of EPS recycling is progressing due to the increase of environmental awareness, strict regulations and technological developments. Increasing demand for the recycled EPS (rEPS) is seen in some developed countries as consumers are becoming more conscious regarding sustainability and environmental impact, increasing the market demand for the recycled products. However, this mindset of consumers is yet to be seen in developing and under- developed countries. The overall efficiency and cost-effectiveness are increasing with the introduction of modern technologies into the recycling process. This is in turn making it attractive to both consumers and businesses. Alongside, recycled EPS is also finding its way into various other applications, such as composite materials and construction. Some fields which are currently utilizing the rEPS most include packaging and automotive interior [31].

Governments are currently enforcing numerous policies and regulations to ensure recycling and reduce environmental negative effects. One of the major policies includes extended producer responsibility (EPR) laws being implemented in various countries worldwide. According to this, the producers or manufacturers are obligated to take the responsibility of end-of-life discarded products and ensure proper recycling or disposal. This is also playing a significant role in EPS recycling [56, 57]. Carbon pricing and emissions trading schemes also impose a financial cost on businesses associated with pollution. These policies encourage recycling of waste materials reducing the environmental impact as high emissions would lead to higher tax which is detrimental to business. Emissions trading schemes include cap and trade which sets a cap on the total amount of greenhouses gas that is allowed to be emitted within a specific jurisdiction. This indicates that the total amount of permitted emissions will be divided into smaller parts. These can be further bought and sold by businesses like stock trades. Therefore, a business exceeding this limit will have to either halt production or buy more permits from other businesses, which involves expenditure and further encourages fewer missions [58].

The opportunities for developing the recycling infrastructure of EPS depend on certain factors that need to be addressed. To improve the EPS recycling rates, it is very important to invest in improved collection and sorting infrastructure to facilitate the efficient collection and processing of EPS wastes. The advancement of more efficient recycling machineries and processes developed through research and developments can help enhance overall efficiency. Ultimately, proper consumer awareness would significantly impact the EPS recycling programs where consumer and businesses' active participation is needed. Apart from this, exploration for new novel applications of rEPS can be game-changing. This would create new markets and increase the demand for rEPS to a great extent. Effective research and development are needed for this purpose, which helps build innovative rEPS products. Moreover, the number of EPS foams sent to landfills for disposal would be reduced significantly, which is highly beneficial for sustainability. Collaboration between various industries would lead to new, multi-functional products that would appeal to consumers. Employing these strategies and utilizing the opportunities, the rEPS industry would grow to establish a sustainable economy to cut carbon emissions and foster businesses significantly.

11 Conclusion

This chapter gives a detailed description of the mechanical recycling process of Expanded Polystyrene foam and environmental benefits together with the business opportunities that may be realized from this process. Therefore, recycling could offer a promising future for EPS recycling provided that some issues associated with contamination, economic efficiency, and technological limitations are addressed. By carrying out more research and development and by also using the right collection, sorting and processing procedures, the recycling rates of EPS waste and the circularity of such waste can be increased. It is possible to state that mechanical EPS recycling for building a more sustainable future is possible as there are successful cases and new emerging technologies. However, it is important to admit the modern issues and limitations that must be solved. Some of the difficulties include contamination, mainly from other plastics and other foreign substances. It is therefore important to note that the feasibility of EPS recycling depends on factors such as market demand, cost of energy and availability of proper facility. Further, technological development is needed to enhance the efficiency of sortation, densification and palletization. However, several challenges have been realized that will affect the recycling of EPS materials in the future. Current research and innovation are resulting in improved technologies in sorting and in the use of EPS in new products. The industry should therefore focus on how to overcome the challenges while at the same time exploiting the opportunities to achieve further improvements in its sustainability. It is also necessary to establish policies and regulations to encourage EPS recycling. Recycling of EPS can be encouraged in the following ways: Governments should come up with incentives to encourage recycling and increase awareness of the same. In this way, through the promotion of EPS recycling, policy makers can help to make the EPS waste more circular and thus decrease its negative impact on the environment. Therefore, it can be concluded that the mechanical recycling of EPS foam is a good chance to prevent the creation of waste, save resources and decrease the level of pollution. Industry can therefore overcome the challenges, build on the opportunities offered by technology and put in place proper policies to ensure that it records further progress towards sustainability.

References

1. Horvath JS (1994) Expanded polystyrene (EPS) geofoam: an introduction to material behavior. Geotext Geomembr 13(4):263–280
2. Simpson A, Rattigan I, Kalavsky E, Parr G (2020) Thermal conductivity and conditioning of grey expanded polystyrene foams. Cell Polym 39:263. https://doi.org/10.1177/0262489320934263
3. Chen W, Hao H, Hughes D, Shi Y, Cui J, Li Z-X (2015) Static and dynamic mechanical properties of expanded polystyrene. Mater Des 69:170–180. https://doi.org/10.1016/j.matdes.2014.12.024
4. Bartolomei SS, da Silva FLF, de Moura EAB, Wiebeck H (2022) Recycling expanded polystyrene with a biodegradable solvent to manufacture 3D printed prototypes and finishing materials for construction. J Polym Environ 30(9):3701–3717. https://doi.org/10.1007/s10924-022-02465-7
5. Khalaj O et al (2020) The experimental investigation of behaviour of expanded polystyrene (EPS). In: IOP conference series: materials science and engineering, institute of physics publishing. https://doi.org/10.1088/1757-899X/723/1/012014
6. Rashid A, Haque M, Islam SMM, Uddin-Labib KMR, Chowdhury P (2024) Breaking boundaries with ceramic matrix composites: a comprehensive overview of materials, manufacturing techniques, transformative applications, recent advancements, and future prospects. Adv Mater Sci Eng 2024(1):2112358. https://doi.org/10.1155/2024/2112358
7. Sayadi AA, Tapia JV, Neitzert TR, Clifton GC (2016) Effects of expanded polystyrene (EPS) particles on fire resistance, thermal conductivity and compressive strength of foamed concrete. Constr Build Mater 112:716–724. https://doi.org/10.1016/j.conbuildmat.2016.02.218
8. Shadrack U, Obele C (2023) A mini-review on expanded polystyrene waste recycling and its applications. World J Adv Eng Technol Sci 8:315–329
9. Rashid A, Showva N-N (2023) Design and fabrication of a biodegradable face shield by using cleaner technologies for the protection of direct splash and airborne pathogens during the COVID-19 pandemic. Clean Eng Technol 13:100615
10. Jolly P (2020) Reducing plastic waste from expanded polystyrene. In: Australian circular economy hub
11. Rashid A, Hoque ME (2022) Polymer nanocomposites for defense applications. In: Hoque ME, Ramar K, Sharif A (eds) Advanced polymer nanocomposites. Woodhead Publishing, New York, pp 373–414
12. Expanded Polystyrene (EPS). British Plastics Federation
13. Rashid A, Showva N-N (2019) Water pollution and its adverse effect on biodiversity in ship breaking area of Bhatiary, Chattogram. Appl Ecol Environ Sci 7(3):96–100
14. Adhikary SK, Ashish DK (2022) Turning waste expanded polystyrene into lightweight aggregate: towards sustainable construction industry. Sci Total Environ 837:852. https://doi.org/10.1016/j.scitotenv.2022.155852
15. Schyns ZOG, Shaver MP (2021) Mechanical recycling of packaging plastics: a review. Macromol Rapid Commun 42(3):2000415. https://doi.org/10.1002/marc.202000415

16. García-Sobrino R, Cortés A, Calderón-Villajos R, Díaz JG, Muñoz M (2023) Novel and accessible physical recycling for expanded polystyrene waste with the use of acetone as a solvent and additive manufacturing (direct ink-write 3D printing). Polymers 15(19):888. https://doi.org/10.3390/polym15193888
17. García M, Duque G, Gracia I, Lucas A, Rodríguez J (2009) Recycling extruded polystyrene by dissolution with suitable solvents. J Mater Cycles Waste Manag 11:2–5. https://doi.org/10.1007/s10163-008-0210-8
18. Dogu O et al (2021) The chemistry of chemical recycling of solid plastic waste via pyrolysis and gasification: state-of-the-art, challenges, and future directions. Prog Energy Combust Sci 84:100901. https://doi.org/10.1016/j.pecs.2020.100901
19. Sriprom W, Sirivallop A, Choodum A, Limsakul W, Wongniramaikul W (2022) Plastic/natural fiber composite based on recycled expanded polystyrene foam waste. Polymers 14(11):241. https://doi.org/10.3390/polym14112241
20. Ragaert K, Delva L, Van Geem K (2017) Mechanical and chemical recycling of solid plastic waste. Waste Manag 69:44. https://doi.org/10.1016/j.wasman.2017.07.044
21. Nathanson S, Jerry A (2021) Solid-waste management. Encyclopedia Britannica
22. Petrella A, Di Mundo R, Notarnicola M (2020) Recycled expanded polystyrene as lightweight aggregate for environmentally sustainable cement conglomerates. Materials 13(4):88. https://doi.org/10.3390/ma13040988
23. Snyder MR (2020) Recycling EPS poses unique sorting challenges. Plastics machinery and manufacturing
24. Li H et al (2022) Expanding plastics recycling technologies: chemical aspects, technology status and challenges. Royal Soc Chem. https://doi.org/10.1039/d2gc02588d
25. Mumbach GD, Bolzan A, Machado RAF (2020) A closed-loop process design for recycling expanded polystyrene waste by dissolution and polymerization. Polymer 209:940. https://doi.org/10.1016/j.polymer.2020.122940
26. Machado NCF et al (2021) Waste-polystyrene foams-derived magnetic carbon material for adsorption and redox supercapacitor applications. J Clean Prod 313:7903. https://doi.org/10.1016/j.jclepro.2021.127903
27. Cunha RS, Mumbach GD, Machado RAF, Bolzan A (2021) A comprehensive investigation of waste expanded polystyrene recycling by dissolution technique combined with nanoprecipitation. Environ Nanotechnol Monit Manag 16:100470. https://doi.org/10.1016/j.enmm.2021.100470
28. Layachi S et al (2023) Effect of incorporating expanded polystyrene beads on thermophysical, mechanical properties and life cycle analysis of lightweight earth blocks. Constr Build Mater 375. https://doi.org/10.1016/j.conbuildmat.2023.130948
29. Taqwatomo G et al (2024) Mechanical properties of recycling mixed waste plastic predicted on pallet application using finite element analysis. In: E3S web of conferences, EDP sciences. https://doi.org/10.1051/e3sconf/202448303019
30. Thakur S, Verma A, Sharma B, Chaudhary J, Tamulevicius S, Thakur VK (2018) Recent developments in recycling of polystyrene based plastics. Elsevier, Amsterdam
31. Assaad JJ, Mikhael C, Hanna R (2022) Recycling of waste expanded polystyrene concrete in lightweight sandwich panels and structural applications. Cleaner Materials 4. https://doi.org/10.1016/j.clema.2022.100095
32. Leão LS, Spini GP, de França MS, Costa EBC (2024) Recycled expanded polystyrene (EPS) as an eco-friendly alternative for sand in rendering mortars. Constr Build Mater 414:135018. https://doi.org/10.1016/j.conbuildmat.2024.135018
33. Abdel-Hakim A, El-Basheer TM, Abd-El-Aziz AM, Afifi M (2021) Acoustic, ultrasonic, mechanical properties and biodegradability of sawdust/recycled expanded polystyrene eco-friendly composites. Polym Test 99:107215
34. Çavuşoğlu A, Karagöz İ (2023) Recycled EPS and epoxy based composite materials: the combination of sustainability and performance. Int J Adv Eng Pure Sci 35(3):379–388. https://doi.org/10.7240/jeps.1285079

35. Reynoso LE, Carrizo-Romero ÁB, Viegas GM, San-Juan GA (2021) Characterization of an alternative thermal insulation material using recycled expanded polystyrene. Constr Build Mater 301:124058
36. Petrella A, Di Mundo R, Notarnicola M (2020) Recycled expanded polystyrene as lightweight aggregate for environmentally sustainable cement conglomerates. Materials 13(4):988. https://doi.org/10.3390/ma13040988
37. Ni X, Wu Z, Zhang W, Lu K, Ding Y, Mao S (2020) Energy utilization of building insulation waste expanded polystyrene: pyrolysis kinetic estimation by a new comprehensive method. Polymers 12(8):81744. https://doi.org/10.3390/POLYM12081744
38. Oladele IO, Okoro CJ, Adelani SO, Agbeboh NI, Betiku OT (2024) Current application of recycled waste plastics as a sustainable materials: a review on availability, processing and application. J Thermoplast Compos Mater 12:08927057241248040
39. Hilal N, Hamah-Sor N, Faraj RH (2021) Development of eco-efficient lightweight self-compacting concrete with high volume of recycled EPS waste materials. Environ Sci Pollut Res 28(36):50028–50051
40. Carvalho CHR, Motta LAC (2019) Study about concrete with recycled expanded polystyrene. Rev IBRACON Estruturas Materiais 12(6):1390–1407. https://doi.org/10.1590/s1983-41952019000600010
41. Kan A, Demirboğa R (2009) A new technique of processing for waste-expanded polystyrene foams as aggregates. J Mater Process Technol 209(6):2994–3000. https://doi.org/10.1016/j.jmatprotec.2008.07.017
42. Akter R, Raja RM (2022) Effectiveness evaluation of shredded waste expanded polystyrene on the properties of binder and asphalt concrete. Adv Civil Eng 2022(1):7429188. https://doi.org/10.1155/2022/7429188
43. Fuesers O, Zumbühl S (2008) The influence of organic solvents on the mechanical properties of alkyd and oil paint. https://api.semanticscholar.org/CorpusID:14812613
44. Sekharan RV, Abraham BT, Thachil ET (2012) Utilization of waste expanded polystyrene: blends with silica-filled natural rubber. Mater Des 40:221–228
45. Gil-Jasso ND et al (2019) Dissolution and recovery of waste expanded polystyrene using alternative essential oils. Fuel 239:611–616. https://doi.org/10.1016/j.fuel.2018.11.055
46. Yeung CWS, Teo JYQ, Loh XJ, Lim JYC (2021) Polyolefins and polystyrene as chemical resources for a sustainable future: challenges, advances, and prospects. ACS Mater Lett 3(12):1660–1676
47. Hattori K (2015) Recycling of expanded polystyrene using natural solvents. In: Recycling materials based on environmentally friendly techniques. InTech. https://doi.org/10.5772/59156
48. de Souza HRA, Dantas TET, Zanghelini GM, Cherubini E, Soares SR (2020) Measuring the environmental performance of a circular system: emergy and LCA approach on a recycle polystyrene system. Sci Total Environ 726:138111
49. Neo ERK, Yeo Z, Low JSC, Goodship V, Debattista K (2022) A review on chemometric techniques with infrared, Raman and laser-induced breakdown spectroscopy for sorting plastic waste in the recycling industry. Resour Conserv Recycl 180:106217
50. Wu X, Li J, Yao L, Xu Z (2020) Auto-sorting commonly recovered plastics from waste household appliances and electronics using near-infrared spectroscopy. J Clean Prod 246:118732
51. Cripps RM (2024) Master gardeners again collect polystyrene. https://www.mymotherlode.com/news/local/3324554/master-gardeners-again-collect-polystyrene.html
52. Casey L (2014) New, free public foam recycling drop-off facility arrives in Texas. https://www.packagingdigest.com/flexible-packaging/new-free-public-foam-recycling-drop-off-facility-arrives-in-texas
53. E. S. G. news (2024) The home depot embraces circular economy: turning waste into opportunity. https://esgnews.com/the-home-depot-embraces-circular-economy-turning-waste-into-opportunity/
54. Wagner TP (2020) Policy instruments to reduce consumption of expanded polystyrene food service ware in the USA. Detritus 9(3):11–26

55. Belblidia F, Gabr MH, Pittman JFT, Rajkumar A (2023) Recycling high impact polystyrene: material properties and reprocessing in a circular economy business model. Progr Rubber Plast Recycl Technol 39(4):343–363
56. Pires A, Martinho G, Ribeiro R, Mota M, Teixeira L (2015) Extended producer responsibility: a differential fee model for promoting sustainable packaging. J Clean Prod 108:343–353
57. Joltreau E (2022) Extended producer responsibility, packaging waste reduction and eco-design. Environ Resour Econ 83(3):527–578
58. Yüksel S, Mikhaylov A (2022) The effect of the carbon tax to minimize emission. In: Clean energy investments for zero emission projects: an analysis on how to reduce the carbon footprint. Springer, pp 1–11

"Industrial Uses of Expanded Polystyrene Foams"

Luisa V. García-Barrera, Victor Varela-Guerrero, and María F. Ballesteros-Rivas

Abstract Currently, the uses and/or applications of Expanded Polystyrene (EPS) are diverse, this is because it is a material that is mostly composed of air (about 98%) and the rest of styrene. It has properties such as lightness, thermal and acoustic insulation, chemical resistance, compression resistance, resistance to humidity, etc. characteristics that make it become a versatile material for use in different industries such as the construction industry, packaging, food, and art industry. This text documents the different uses of EPS in each of these industries, as well as research and studies related to these applications and future perspectives that exist for this polymer.

Keywords Expanded polystyrene (EPS) · EPS industry · EPS applications · Recycling

1 Introduction

Nowadays, the abuse of Expanded Polystyrene (EPS) consumption has generated significant amounts of pollution worldwide. We observe this consumption in our daily lives due to the qualities and properties that this material offers us, such as: resistance, lightness [1], acoustic insulation, resistance to humidity, thermal resistance, chemical inertia, durability [2], qualities that make the applications and uses of EPS convenient for different industries, this according to the final use that is wanted to be given to the material.

L. V. García-Barrera · V. Varela-Guerrero (✉) · M. F. Ballesteros-Rivas (✉)
Universidad Autónoma del Estado de México. Facultad de Química, Toluca, México, México
e-mail: vvarelag@uaemex.mx

M. F. Ballesteros-Rivas
e-mail: mfballesterosr@uaemex.mx

Centro Conjunto de Investigación en Química Sustentable UAEM-UNAM, Toluca de Lerdo, Mexico

A. Khan et al. (eds.), *Uses and Products of Recycled Expanded Polystyrene Foam Wastes*, Sustainable Materials and Technology,
https://doi.org/10.1007/978-981-96-5618-9_2

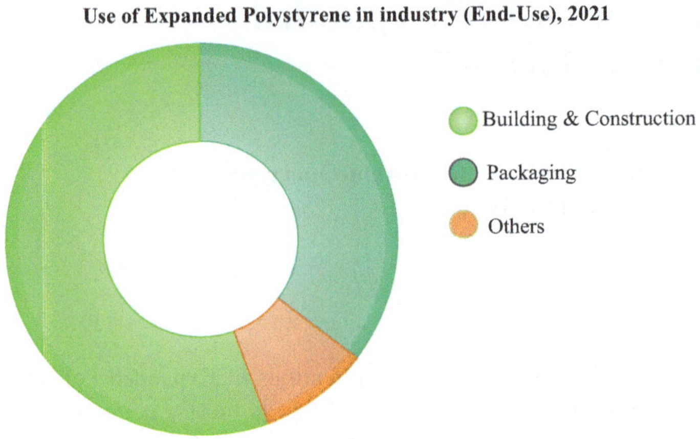

Graph 1 Use of expanded polystyrene in industry (end-use), 2021. *Note* Information adapted from the source Acumen Research And Consulting and Advisory [4]

In 2016, the global production of Polystyrene and EPS was around 14.7 and 6.6 MMT (Million Metric Tons) [1]. Currently, the Expanded Polystyrene market is estimated at 12.46 million tons in 2024 and is expected to reach 14.48 million tons in 2029 [3].

The uses of Expanded Polystyrene can be observed in the industry, in fields such as construction, industrial packaging, food, and others (See Graph 1) directly or also as additives for specific applications.

2 Construction Industry

Expanded polystyrene has been introduced in the construction industry for the construction of buildings and residences, thus forming part of the materials for their construction. High-density concrete and lightweight concrete are popular for construction due to their low cost. Some uses require a high-density material, but also depending on the application, materials with a lower density are required [5]. Examples of uses/applications of EPS in this industry are shown in Fig. 1.

2.1 Use of EPS as Thermal Insulation

In this industry, there are a variety of materials for building constructions, however, there are other challenges such as energy consumption in these, due to the use of heaters for temperature control, however, these are not enough in addition to the fact that there are energy losses [6], which leads to a demand for energy to cover this need,

Fig. 1 Common applications of EPS in the construction industry

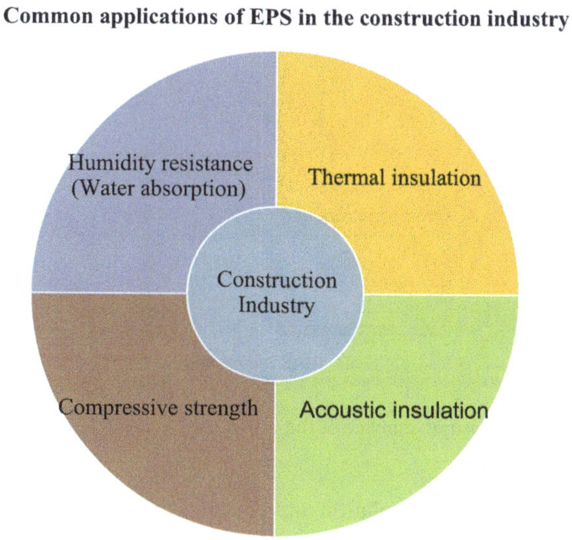

consequently today the construction industry seeks thermal insulation alternatives with good thermal efficiency at a lower cost, and one of these alternatives is the use of Expanded Polystyrene as a thermal insulation material.

By definition, thermal insulators are materials or a combination of materials that are used to cause resistance to heat flow [7]. As is well known, Expanded Polystyrene are polystyrene beads that are usually expanded with pentane C_5H_{12} so that these beads join together during the expansion process [8]. Due to this structure, this material has the property of having a low density and, together with a high thermal barrier, they allow obtaining lightweight insulation blocks for use as construction material.

Expanded polystyrene has a thermal conductivity of 0.030 and 0.040 W/(mK) although this can vary due to factors such as humidity, temperature and density of the mass [6].

Although concrete is the most widely used material for construction due to its strength and durability [5], there is currently research to improve construction materials. Sariisik and Sariisik [9], developed a new method for producing lightweight concrete and EPS composite blocks, with pumice aggregates. The study revealed good thermal and acoustic insulation (thermal conductivity of 0.33 W/mK and 60 dB for acoustic insulation). In addition, this new production of blocks has the advantage that by incorporating EPS foam into lightweight concrete with pumice, the unit weight along with the cost of these insulation blocks is lower compared to other construction materials, making it an economical material.

On the other hand, the study carried out by Demirel [10] used pumice and EPS to manufacture insulation blocks, focusing on determining the optimal thickness for minimum heat transfer, finding the value of approximately 40 and 20 mm, obtaining

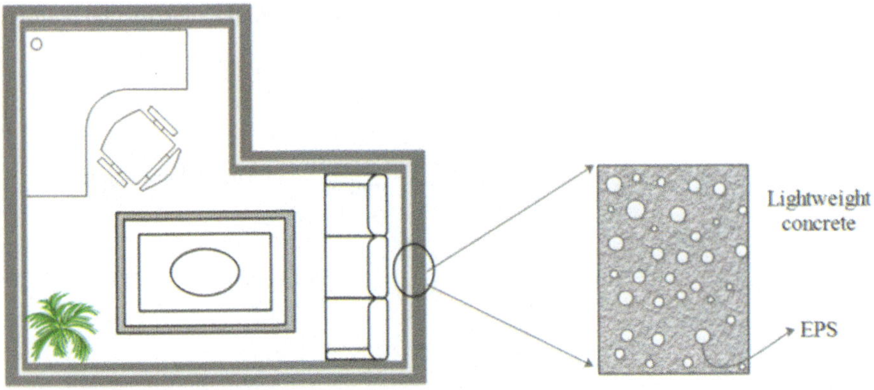

Fig. 2 Illustrative image of the use of EPS as an additive in lightweight concrete

good thermal insulation and compressive strength, with the additional characteristic that the EPS block with pumice is lighter.

2.2 *Use of EPS as an Acoustic Insulator*

Another example of this is noise insulation or acoustic insulation (See Fig. 2), in which the main function is to control the propagation of sound, that is, to reduce the transmission of noise between one place and another [11]. Recent research has focused on the reuse of EPS to replace gravel in the manufacture of lightweight cement [12], an example of this is the work carried out by Santos et al. [13], in which Expanded Polystyrene beads were added to concrete, discovering that this lightweight additive offers a 50% reduction in the transmission of sound vibrations, creating a good material for noise insulation.

2.3 *Use of EPS in Humidity Resistance*

Expanded Polystyrene as an additive in concrete also plays an important role in water absorption, since it is an important factor for the durability of the material. Studies have been carried out on the behavior of EPS/concrete, in which different behaviors are observed, this according to the quantity or concentration of EPS in the mixture with the concrete [14].

On the one hand, it has been reported that lightweight concrete with the addition of Expanded Polystyrene beads reduces water adsorption compared to a typical sand/concrete mixture, an example of this is the case of Abdullah Imtiaz et al. [15], who in their research reported a decrease in water absorption from 6.90% (Concrete and sand mixture) to 3.65%, 3.80% and 4.12% (Lightweight concrete with EPS aggregates), this is attributed to the closed cell structure of EPS.

On the other hand, an increase in water absorption is reported [5] when adding EPS to concrete, as it is well known the absorption process depends on the porosity of the material, to which adding EPS beads to concrete improves absorption by increasing the porosity of the EPS/concrete mixture, Adhikary and Ashish [14] explain that this increase in absorption is due to the poor adhesion between concrete and Expanded Polystyrene, in addition to the existence of spaces between these two materials causing a volume of air to be trapped.

However, research has managed to reduce this increase in adsorption by implementing microsilica and nanosilica creating optimal adhesion between the Expanded Polystyrene beads and the concrete [16].

2.4 Use of EPS for Compressive Strength as an Aggregate in Concrete

There are other parameters such as compressive strength, a quality to determine the quality of concrete, as defined it is the capacity of the material to resist tension without breaking [17], according to ACI-213R-2014, structural concrete must have a compressive strength of ≥17 MPa and a density between 1350–1900 kg/m^3. There are studies on how Expanded Polystyrene influences concrete for the evaluation of this mechanical property, an example of this is cellular concrete or expanded concrete, whose main characteristic is the integration of macroscopic voids [5], de Souza et al. [5] obtained a 93% resistance in three days, attributing this result to the number of voids in the concrete structure, a higher cement content, as well as the low density of the EPS granules.

However, other investigations have had results in which the compressive strength is compromised by a low cement content in the concrete and a lack of natural coarse aggregate in the mixture [15].

2.5 Use of Recycled Expanded Polystyrene in the Construction Industry

Currently, the recycling of Expanded Polystyrene has become a topic of interest due to its high consumption, causing damage to the environment, in addition to the fact that its useful life is short and takes years to decompose. Due to this, in the field of construction and civil engineering, materials that are friendly to the environment are sought.

Nowadays, the incorporation of recycled Expanded Polystyrene in concrete has been sought. This practice leads to the recycling of Expanded Polystyrene, since a second useful life is given to Expanded Polystyrene as part of the construction material, thus avoiding the use of more raw material. In addition, it offers benefits such as reducing the consumption of raw materials by making EPS part of the construction material, since in the construction industry there is a considerable consumption of raw materials in addition to the fact that there is a considerable generation of waste during the process [18].

With this objective, Expanded Polystyrene waste has currently been incorporated into concrete production. In the manufacture of concrete with commercial EPS vs recycled EPS, their properties vary slightly, however the behavior between one material and another is almost similar/analogous [14].

Researchers have taken on the task of investigating these properties using recycled Expanded Polystyrene, to evaluate the feasibility of the material to be used in constructions, as well as to examine possible improvements so that this practice is encouraged. An example of this is Zaragoza et al. [19].

We know that today construction materials are essential for homes and buildings, however, to have a sustainable construction industry, it is necessary to work and implement materials that meet essential requirements to reduce costs and energy in the construction of buildings and homes. As previously mentioned, recycling expanded polystyrene is a good practice since it promotes this practice, thus developing ecological constructions and favors caring for the environment.

3 Packaging Industry

Another use or application of Expanded Polystyrene is in the packaging industry, due to its properties such as resistance to compression (ideal characteristic for stackable packaging goods), moisture resistance, lightness, chemical resistance, durability [20], modeling capacity (shapes) [21], among others, makes this material widely used in this sector and therefore the second industry (See Graph 1) where a significant amount of Expanded Polystyrene is consumed [4, 22].

There is a wide variety of products in which packaging with Expanded Polystyrene is required, examples of them are: household appliances, components and electronic material, furniture, tools, chemical products, pharmaceuticals, perfumery, cosmetics, etc., items that are fragile in nature [4], or due to their components they require more specialized (more careful) storage and transportation (See Fig. 3).

This application is usually seen in the storage of items, as well as their transportation, since it is an excellent cushioning material, that is, it provides impact protection to different products [23]. In addition to this protection, this characteristic of EPS prevents the merchandise from being damaged by vibrations during handling and/or transportation.

In the packaging industry, one of the important factors when manufacturing this type of material is the design, since this will determine the protection efficiency it offers [21]. Two types of polystyrene foam are usually used in packaging, the first is

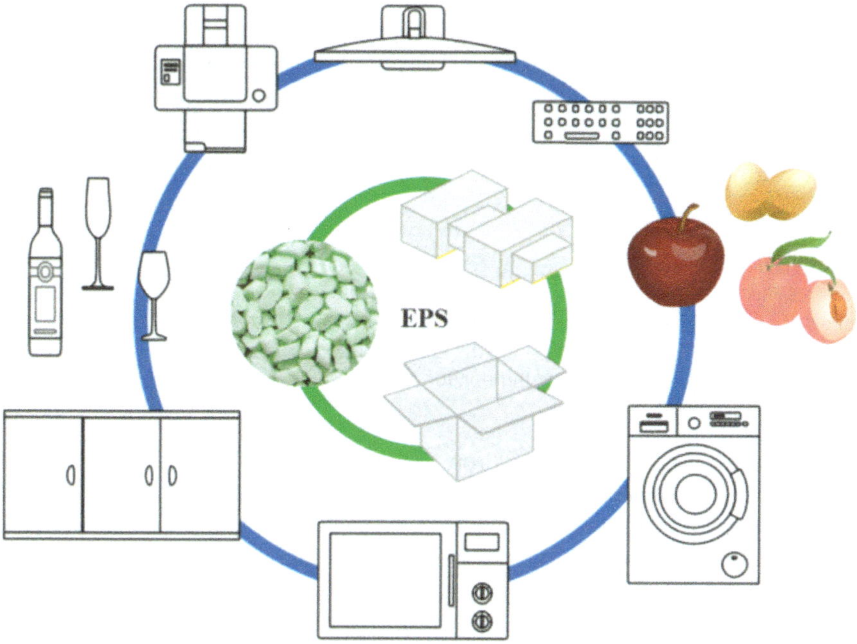

Fig. 3 Uses of EPS in the packaging industry

open-cell rigid expanded foam and the second is closed-cell flexible expanded foam [21] (See Fig. 4), depending on the product, one or the other is used.

As mentioned above, EPS has the quality of being modeled in a variety of shapes and sizes, these modeled according to the consumer's need. These designs are key to fulfilling the function of avoiding vibrations and/or impacts of the product, since parameters such as the cushioning area are evaluated [23] and even studies are carried

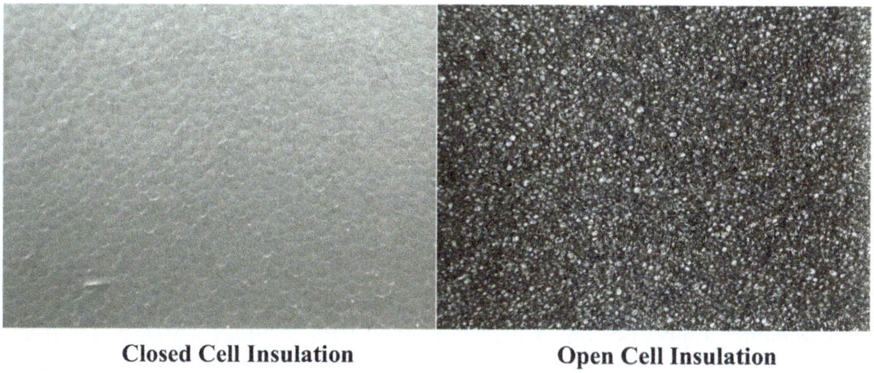

Fig. 4 Closed cell insulation and open cell insulation

Table 1 Examples of packaging types

Packaging shape	Special protections
Separators and supports	Returnable packaging
Closed packaging	Industrial packaging
Boxes for food products	Thermal boxes
	Ice cream container
	Frozen desserts container
	Fillet containers
	Botellpack®
	MastroPack®

Source MASTROPOR® [24]
Note Information adapted

out in the design process such as drop tests to see the faults and weaknesses of the design, for example Kassim et al. determined parameters such as Rib, Thinkness, Layout and Sice for sustainable EPS packaging [21], thus existing a variety of configurations of shapes and types of packaging, an example of this is shown in Table 1.

3.1 Future Prospects for the Use of EPS in the Packaging Industry

Nowadays, the use of Expanded Polystyrene has become an environmental problem, due to its notable use in daily life. EPS being a non-biodegradable material, recycling this polymer has become a challenge, since the objective when recycling this material is to reduce the volume of EPS, as well as reduce storage and transportation costs, due to the characteristic of being a bulky material [25]. Expanded Polystyrene in the packaging industry has a short useful life, since it ends up as waste ending up in a landfill due to improper disposal [26].

For this reason, studies have been developed for the possible replacement of this material in packaging. Joyal et al. [26] worked on a biocomposite as a possible substitute for traditional EPS packaging, in which they used sawdust and mycelium fibers as a binding agent to form a composite, to which the results obtained by their research highlighted a dense and compact material, better thermal stability, less emission of toxic agents in a combustion process, and of course the nature of being a biodegradable material.

On the other hand, Salini et al. [27], implemented the recycling of Expanded Polystyrene for possible packaging applications. In their study, they synthesized a biodegradable composite film, which contained EPS residues, cassava starch and epoxidized neem oil, obtaining good mechanical and thermal stability, low water absorption, good percentage of elongation, biodegradability is shown at 180 days.

4 Food Industry

The use of Expanded Polystyrene is also present in food packaging, due to the food safety offered by EPS [20], such as in the packaging of fish and seafood, meat and poultry products, fruits and vegetables, where EPS protects the products from impacts, dairy products, and EPS is also present in the fast food industry [28].

These types of Expanded Polystyrene containers are regulated by the Food and Drug Administration (US FDA) to ensure public safety and food, medicines, as well as medical devices, cosmetics for safe consumption [29], there are also regulations such as: (EC) No. 1935/2004, (EC) No. 2022/1616 and (EU) 2016/1416 that aim to establish certain criteria to ensure the safe consumption of food products that are in contact with certain polymers, such as EPS, including the manufacturing of these containers [20].

However, the consumption of EPS in this sector also plays an important role in the contamination of the environment by these wastes. The recycling of Expanded Polystyrene in the food industry is complicated since it can include certain contaminants and impurities, causing it not to be used in the packaging of food or food that is in contact with these recycled plastics [28]. An example of this is in China and Europe, where recycled plastics are prohibited from being used to be in contact with food [28].

Due to this impediment, it is important to implement materials that are more environmentally friendly. Velasco et al. [30], developed a compostable container based on potato starch, seeking to keep chicken meat fresh, resulting in good mechanical properties, efficient liquid retention for the purpose of keeping chicken meat fresh, and by adding nisin it obtained an antimicrobial effect.

Another similar example is that of Amaraweera [31], who together with his collaborators developed a method for the synthesis of foam based on cassava starch, a promising material for packaging applications.

In conclusion, the use of Expanded Polystyrene in disposable materials (plates, trays, cups) for the food industry is important to have sustainable and environmentally friendly alternatives, since in this industry EPS is handled for single use, causing increasing pollution due to the use of this material.

5 Uses of Expanded Polystyrene in the Aesthetic Industry

Another important field to consider is the use of Expanded Polystyrene in the aesthetic industry, calling this type of industry because EPS is used in an artistic, creative way or in the form of crafts and decoration.

In this application there are two postulates, some consider that the use of Expanded Polystyrene in aesthetics leads to significant contamination due to the use of virgin EPS, and on the other hand there is a tendency toward the use of EPS in this field, but for the purpose of recycling this material [32].

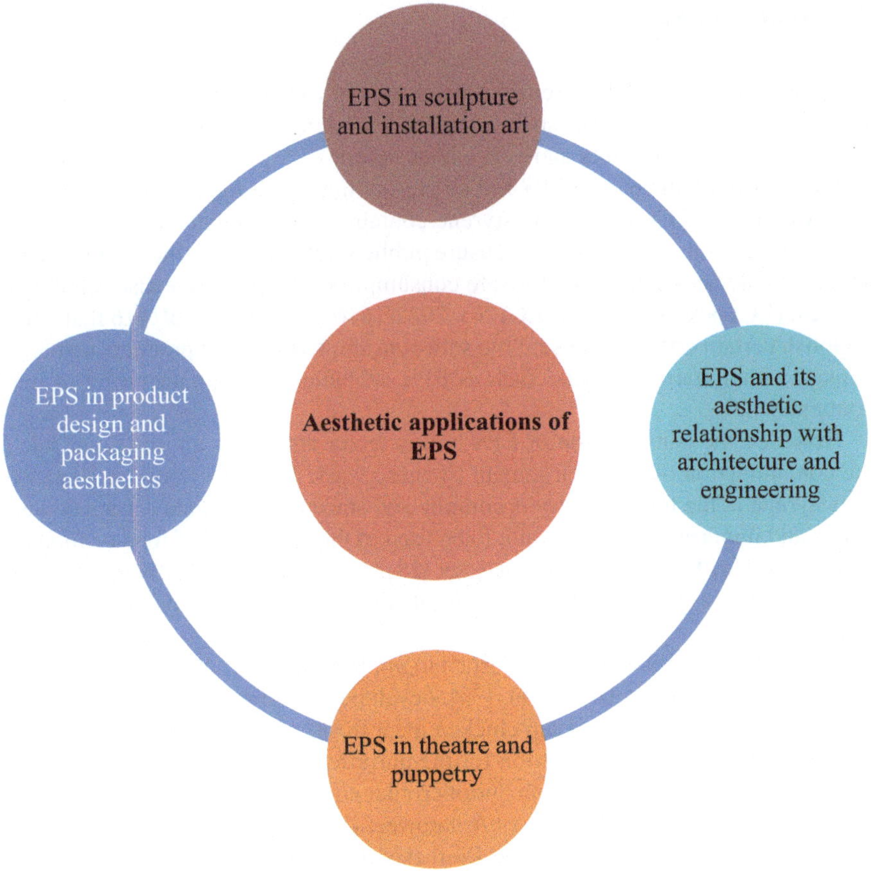

Fig. 5 Aesthetic applications of EPS [32]. *Note* Information adapted

Some examples of the use of EPS are shown in Fig. 5

Taking into account the above, it is important in this field to implement the recycling of Expanded Polystyrene to make this type of products for artistic purposes, so that this practice is sustainable and helps with the environmental problems caused by Expanded Polystyrene waste.

6 Conclusion

As we have seen, there is a wide use of Expanded Polystyrene, ranging from the construction, packaging, food, aesthetics and other applications that we can mention such as: protective helmets, life jackets, fillings for automobile components, etc. Due to the characteristics and properties of Expanded Polystyrene, a variety of items can be manufactured for different purposes, creating advantages over some other materials, providing versatility in different industries due to the use of EPS. However, excessive use in some fields leads to significant contamination due to the short useful life of Expanded Polystyrene, as in the case of disposable packaging, in turn due to inadequate disposal and the low recyclability of this material, so it is important to implement and carry out the practice of recycling, since as mentioned above there are methods and alternatives in each of the industries for the total or partial replacement of this material, thus promoting the development of sustainable materials and consequently the care of the environment.

References

1. Hidalgo-Crespo J, Moreira CM, Jervis FX, Soto M, Amaya JL, Banguera L (2022) Circular economy of expanded polystyrene container production: environmental benefits of household waste recycling considering renewable energies. Energy Rep 8:306–311. https://doi.org/10.1016/j.egyr.2022.01.071
2. ECOPLAS (2020) The contributions of EPS EXPANDED POLYSTYRENE to the circular economy. [Online]. Available: www.ecoplas.org.ar
3. Mordor Intelligence (2024) Expanded polystyrene market size and share analysis growth trends and forecasts (2024–2029). Tamaño del mercado de poliestireno expandido y análisis de participación tendencias de crecimiento y pronósticos (2024–2029). https://www.mordorintelligence.com/es/industry-reports/global-expanded-polystyrene-eps-market-industry. Accessed 9 Sept 2024
4. Acumen Research And Consulting (ARC) (2022) Expanded polystyrene market size—global industry, share, analysis, trends and forecast 2022–2030. [Online]. Available: https://www.acumenresearchandconsulting.com/expanded-polystyrene-market
5. de Souza TB, Medeiros MHF, Araújo FWC, de Melo Neto AA (2023) The influence of expanded polystyrene granules on the properties of foam concrete. Mater Struct Constr 56(1). https://doi.org/10.1617/s11527-023-02109-9
6. Abu-Jdayil B, Mourad AH, Hittini W, Hassan M, Hameedi S (2019) Traditional, state-of-the-art and renewable thermal building insulation materials: an overview. Constr Build Mater 214:709–735. https://doi.org/10.1016/j.conbuildmat.2019.04.102
7. Conuee México (2009) Benefits of thermal insulation in the industry, pp. 1–52, [Online]. Available: https://www.gob.mx/cms/uploads/attachment/file/93853/aislamiento.pdf
8. Pérez García N, Garnica Anguas P, Pérez Salazar A, Juárez Hidalgo R, Castro Mondragón F (2016) Evaluation of the mechanical properties of expanded polystyrene. Inst Mex del Transp 476:1–79. [Online]. Available: https://imt.mx/archivos/Publicaciones/PublicacionTecnica/pt476.pdf
9. Sariisik A, Sariisik G (2012) New production process for insulation blocks composed of EPS and lightweight concrete containing pumice aggregate. Mater Struct Constr 45(9):1345–1357. https://doi.org/10.1617/s11527-012-9836-z
10. Demirel B (2013) Optimization of the composite brick composed of expanded polystyrene and pumice blocks. Constr Build Mater 40:306–313. https://doi.org/10.1016/j.conbuildmat.2012.11.008
11. Texsa (2009) Acoustic insulation systems for new construction and renovation 23(222):55–85. https://doi.org/10.3989/ic.1970.v23.i222.3547
12. Ducman V, Mirtič B (2009) The applicability of different waste materials for the production of lightweight aggregates. Waste Manag 29(8):2361–2368. https://doi.org/10.1016/j.wasman.2009.02.013

13. Santos LL, Pontes ÍS, Bastos LP, de Melo GV, Barata M (2020) Acoustic performance of social housings in Brazil: assessment of lightweight expanded polystyrene concrete as resilient subfloor. J Build Eng 41(July):2021. https://doi.org/10.1016/j.jobe.2021.102442
14. Adhikary SK, Ashish DK (2022) Turning waste expanded polystyrene into lightweight aggregate: towards sustainable construction industry. Sci Total Environ 837:155852. https://doi.org/10.1016/j.scitotenv.2022.155852
15. Imtiaz A, Sharif MB, Irfan-ul-Hassan M, Ghafoor MT, Hasan S (2024) Effect of expanded polystyrene beads on mechanical, thermal, and acoustic properties of lightweight concrete. Arab J Sci Eng. https://doi.org/10.1007/s13369-024-09026-1
16. Fathi M, Yousefipour A, Hematpoury Farokhy E (2017) Mechanical and physical properties of expanded polystyrene structural concretes containing micro-silica and nano-silica. Constr Build Mater 136:590–597. https://doi.org/10.1016/j.conbuildmat.2017.01.040
17. León Consuegra L, Rodríguez García C (2011) Factors influencing the compressive strength of concrete. State of the art. J Archit Eng 16(3):2, ISSN-e 1990–8830. [Online]. Available: https://dialnet.unirioja.es/servlet/articulo?codigo=8784281&info=resumen&idioma=ENG, https://dialnet.unirioja.es/servlet/articulo?codigo=8784281&info=resumen&idioma=SPA
18. Scioti A, Fatiguso F (2024) Performance characterization and evaluation of innovative cement mortars and concretes made with recycled EPS. Appl Sci 14(4). https://doi.org/10.3390/app14041583
19. Zaragoza-Benzal A, Ferrández D, Atanes-Sánchez E, Morón C (2023) New lightened plaster material with dissolved recycled expanded polystyrene and end-of-life tyres fibres for building prefabricated industry. Case Stud Constr Mater 18. https://doi.org/10.1016/j.cscm.2023.e02178
20. ANAPE (2024) Packaging. Properties of EPS for Packaging. https://anape.es/envase-embalaje/#propiedades
21. Kassim N et al (2023) Sustainable packaging design for molded expanded polystyrene cushion. Materials (Basel) 16(4). https://doi.org/10.3390/ma16041723
22. CHEMANALYST (2023) Expanded polystyrene market analysis: industry market size, plan capacity, production, operating efficiency, demand & supply, end-user industries, sales channel, regional demand, company share, manufacturing process, 2015–2032. https://www.chemanalyst.com/industry-report/expanded-polystyrene-market-2967
23. Lye SW, Lee SG, Tor SB (1998) A parametric study of the shock characteristics of expandable polystyrene foam protective packaging. Polym Eng Sci 38(4):558–565. https://doi.org/10.1002/pen.10218
24. Mastropor SA (2022) Industrial packaging line. Expanded polystyrene (EPS) Expanded polypropylene (EPP). Argentina 1–15
25. Samper MD, Garcia-Sanoguera D, Parres F, López J (2010) Recycling of expanded polystyrene from packaging © smithers rapra technology. Prog Rubber Plast Recycl Technol 26(2):83–92
26. Jose J, Uvais KN, Sreenadh TS, Deepak AV, Rejeesh CR (2021) Investigations into the development of a mycelium biocomposite to substitute polystyrene in packaging applications. Arab J Sci Eng 46(3):2975–2984. https://doi.org/10.1007/s13369-020-05247-2
27. Salini NG, Resmi BG, Antony R (2021) Biodegradable composites of waste expanded polystyrene with modified neem oil for packaging applications. J Elastomers Plast 53(8):975–991. https://doi.org/10.1177/0095244321l006164
28. Song XC, Lin QB, Zhang YC, Li Z, Zeng Y, Chen ZF (2018) Rapid classification of virgin and recycled EPS containers by Fourier transform infrared spectroscopy and chemometrics. Food Addit Contam Part A Chem Anal Control Expo Risk Assess 35(11):2220–2229. https://doi.org/10.1080/19440049.2018.1515502
29. U.S. FOOD & DRUG ADMINISTRATION (2024) What does FDA regulate? A guide for health professionals. https://www.fda.gov/about-fda/what-we-do/que-regula-la-fda
30. Velasco V, Sepúlveda E, Williams P, Rodríguez-Llamazares S, Gutiérrez C, Valderrama N (2022) Starch-based composite foam for chicken meat packaging. J Food Sci Technol 59(12):4594–4602. https://doi.org/10.1007/s13197-022-05538-6
31. Amaraweera SM et al (2022) Preparation and characterization of dual-modified cassava starch-based biodegradable foams for sustainable packaging applications. ACS Omega 7(23):19579–19590. https://doi.org/10.1021/acsomega.2c01292

32. Awodi E, Adewumi KC (2024) Exploring the aesthetic applications of expanded polystyrene: an interdisciplinary review an overview of expanded polystyrene. African J Inter/Multidisciplinary Stud 6(1):1–15

New Packaging Products from Recycled Expanded Polystyrene Foams

Uplabdhi Tyagi and Anil Kumar Sarma

Abstract This chapter elucidates techniques for producing packaging foams from recycled expanded polystyrene foams (EPS) for multiple utility. Advantages of EPS over other packaging materials and their environmental impacts threatening sustainability have also been discussed. Method for recycling the EPS viz., mechanical, chemical and thermal with specific advancement in technology and recent publications have also been discussed. Biodegradable EPS and modern approach to use biological resources for EPS degradation and microwave assisted EPS degradation are some thrust areas of research. The primary focus of all recycling processes is however, the resource recovery with minimum environmental impacts.

Keywords Expanded polystyrene foams · Degradation complexity · Recycling technologies · Packaging applications

1 Introduction

Expanded polystyrene (EPS) foams are essential in the modern world of packaging. Their special properties including light weight, high insulation capability, and low cost make them valuable for application in packaging of food and beverages, consumer electronics, construction items, and medical devices. Still, waste EPS remains a problem because of widespread use, as it degrades at such a slow natural rate that it becomes not useful within hundreds of years. Since EPS constitutes the majority of polystyrene plastic, its lifetime would persist for so many hundred years in environments that fill landfills, causing marine pollution; hence, recycling waste

U. Tyagi
University School of Chemical Technology, Guru Gobind Singh Indraprastha University, Delhi, India

A. K. Sarma (✉)
Chemical Conversion Division, Sardar Swaran Singh National Institute of Bio-Energy (An Autonomous Institute of MNRE, Government of India), Kapurthala, Punjab, India
e-mail: anil.sarma16@gov.in

A. Khan et al. (eds.), *Uses and Products of Recycled Expanded Polystyrene Foam Wastes*, Sustainable Materials and Technology,
https://doi.org/10.1007/978-981-96-5618-9_3

EPS is very crucial. The aim of recycling EPS is to transform the material back into reusable forms, creating a sustainable solution that has preserved its valuable properties while reducing impacts on the environment. The most efficient means of recycling EPS include its mechanical, chemical, as well as thermal process, and one of the popular methods of mechanical recycling encompasses grinding and melting the EPS with the help of a pellet for molding new products. Although this technique is energy saving and cheaper, it fails in cases of contaminated EPS [1]. In contrast, chemical recycling breaks down at a molecular level and offers the chance of recovery of all the basic components of EPS toward quality recycled material. This approach is, however more involved and costly but can involve contaminated EPS, thus affording a comprehensive recycling process. Thermal recycling, on the other hand, breaks down the EPS at high temperatures and captures its energy-rich content to generate power. This has a major disadvantage that emissions generated in this process must be strictly controlled in order not to cause pollution. Recent developments include hybrid and advanced recycling methods that bring elements of both mechanical and chemical recycling together for enhanced efficiency, quality, and reduced environmental impact. New products in terms of recycled EPS arise along with the advancement of the technology of recycling EPS. With the help of EPS, waste is prevented from furthering the building of the circular economy. Lower demands on pristine materials that involve the adoption of recycled EPS for packaging reduce the ecological footprint associated with packaging industries. Moreover, the recycling of EPS into packaging material shows very good results concerning thermal insulation, shock absorption, and durable properties in comparison to pristine EPS [2]. That means that recycled EPS may be used in various forms for packaging applications including molded packaging for electronics, sheets of insulation for construction applications, and even for the more fragile custom packing requirements. Apart from the functional properties, the recycling of EPS packaging material has significant environmental advantages. Recycling EPS saves the industry from extracting petroleum resources and hence reduces the carbon footprint of the whole industry. For example, researches have indicated that EPS recycling can reduce greenhouse gas emissions by up to 50% compared to pristine polystyrene production. In addition, recycled EPS offers a more sustainable option in that it can be cycled several times without major loss in performance. It is also aligned with the growing trend toward a circular economy, wherein materials are reused and recycled and will not cause any kind of adverse effects on the environment. However, despite these advancements, there are a few challenges that packaging is faced with when it is being extensively used in recycling as EPS. The foremost of those is contamination. In the case of food packaging EPS, grease or other contaminants in foodstuff disrupt the recycling process. This type of contamination also lowers the grade of recycled EPS and increases the processing cost since very high purification is required. The other minor challenges include the lack of already put in place infrastructures for recycling and proper collection systems specifically designed to accommodate EPS. Since EPS is voluminous, many recycling plants are not set up to process EPS since it is not in demand in the market thus recycling rates are limited. One negative effect of recycling on product quality is contamination with food residues or other materials. Also, economically, there exists an aspect where collecting, processing, and transporting EPS to be recycled might be very costly than pristine materials and such economic inequality somewhat hinders recycled EPS in penetrating the market. Last but not the least, technology to effectively convert used EPS into high-quality material is still under development that would sometimes limit the range of applications for recycled EPS [3]. Nevertheless, research work continues and technological advancements can already provide answers to these dilemmas with improved recycling methods minimizing the effects of contaminants, ensuring quality end product output.

Therefore, the main focus of this chapter is to examine the journey of EPS from waste to resource, outlining the current technologies, challenges, and future prospects of recycled EPS in packaging applications. The subsequent sections will describe recycling processes of EPS in further detail, from established ones to emerging ones, the technical properties of the used recycled EPS, and retained in packaging applications. Further applicability in other industries may lead toward contribution toward sustainable packaging practices. This innovation, coupled with favorable policies and public awareness, is considered necessary for sustainable production during the shift of industries toward complete potential utilization of recycled EPS foams in packaging applications. From this analysis, it can be seen that there is a possibility of solving environmental issues without compromising on the functional demands of modern packaging applications through the usage of recycled EPS.

2 The Journey of EPS from Waste to Resource

The decades have seen transformations in how EPS foam is perceived, and its role in it. It was discovered in 1949, by D. Dow, in Germany. It soon became one of the most haled material for use in the 1950s, much to be sought after because it was light in weight with excellent insulating properties most especially fragile items. It was only from the 1960s that EPS started to gain usage as a building material to erect energy-efficient structures. Its usage gained grounds in the 1970s among the food service in disposable cups and take-out containers. However, it also gave rise to the escalating environmental issues confronted by the United States of America. The EPA began feeling concerned over the waste concern of EPS, just as with other foams. The EPS exhibits high resistance to impact, superior heat insulation, and resistance to moisture properties, but other material alternatives have advantages that prove sustainable at the cost of durability, tolerance for moisture, and weight (Table 1). Because of its slow degradation rate (Fig. 1), it can accumulate in natural environments and degrade into harmful microplastics that affect the wildlife and penetrate the food chain. In addition, the production and burning of EPS contribute to greenhouse gases, which raise concern about air quality and global warming. Evidently, the perspective has shifted, wherein EPS is becoming far more important as a resource rather than waste [4]. There are new applications of recyclable EPS in insulation and other new packaging, which is also oriented toward the principle of a circular economy, seeking to enhance resource use efficiency and minimize waste. This transformation is further

Table 1 Performance metrics of EPS in comparison to alternative packaging materials

Material	Properties	Environmental impact	Cost implications	Ref.
Expanded polystyrene (EPS)	Lightweight (density: 10–50 kg/m^3), excellent thermal insulation (R-value: 3.6–4.2 per inch), high impact resistance	Non-biodegradable (decomposes over hundreds of years), contributes approximately 30% to global plastic waste, significant marine pollution risk	Low-cost production (~\$0.80 to \$1.50 per kg) due to efficient manufacturing processes	[6]
Polyethylene (PE)	Flexible (elongation at break: 300–800%), moisture-resistant, tensile strength: ~20–30 MPa	Recyclable (~30% recycling rate), energy-intensive (2–4 MJ/kg for recycling), contributes to microplastic pollution	Generally low (\$0.70 to \$1.20 per kg), varies with type (e.g., LDPE, HDPE)	[7]
Polypropylene (PP)	Rigid, high chemical resistance (can withstand temperatures up to 100 °C), tensile strength: ~30–50 MPa	Moderate recycling rates (~10% recycling rate), fossil fuel-based, low biodegradability	Higher than EPS (\$1.50 to \$2.50 per kg), cost affected by crude oil prices	[8]
Biodegradable plastics	Compostable (biodegrades in 90–180 days in industrial composting), derived from renewable resources (e.g., corn starch, PLA)	Reduces landfill impact, however, not widely available; requires specific conditions for composting	Higher production costs (\$2.00 to \$4.00 per kg), limited by availability and performance	[9]
Paper-based packaging	Biodegradable, renewable, tensile strength: ~40–100 MPa, lightweight (density: 300–800 kg/m^3)	Lower carbon footprint (~0.5–1.2 kg CO_2 per kg produced), but concerns over deforestation and water use in production	Moderate (\$1.00 to \$2.00 per kg), dependent on sourcing practices and certifications	[10]

supported by rising awareness among the people and change in policies of curtailing the use of single-use products of EPS in the commercial sector. Society can have a sustainable future in the management of EPS by considering the environmental impacts associated with EPS waste and embracing its potential for recycling and reuse. The 1990s witnessed the introduction of recycling facilities that reduced the amount of EPS to be put into landfills. When the cities began banning the use of single-use products made from EPS in the early 2000s, the industry started looking for ways to be environmentally friendly. It came up with recycling schemes and created biodegradable alternatives [5]. To this date, EPS is used but with the strongest motivation to find a more ecological way of making the plastic. This history calls for the balancing of those useful properties of EPS as an environmental need.

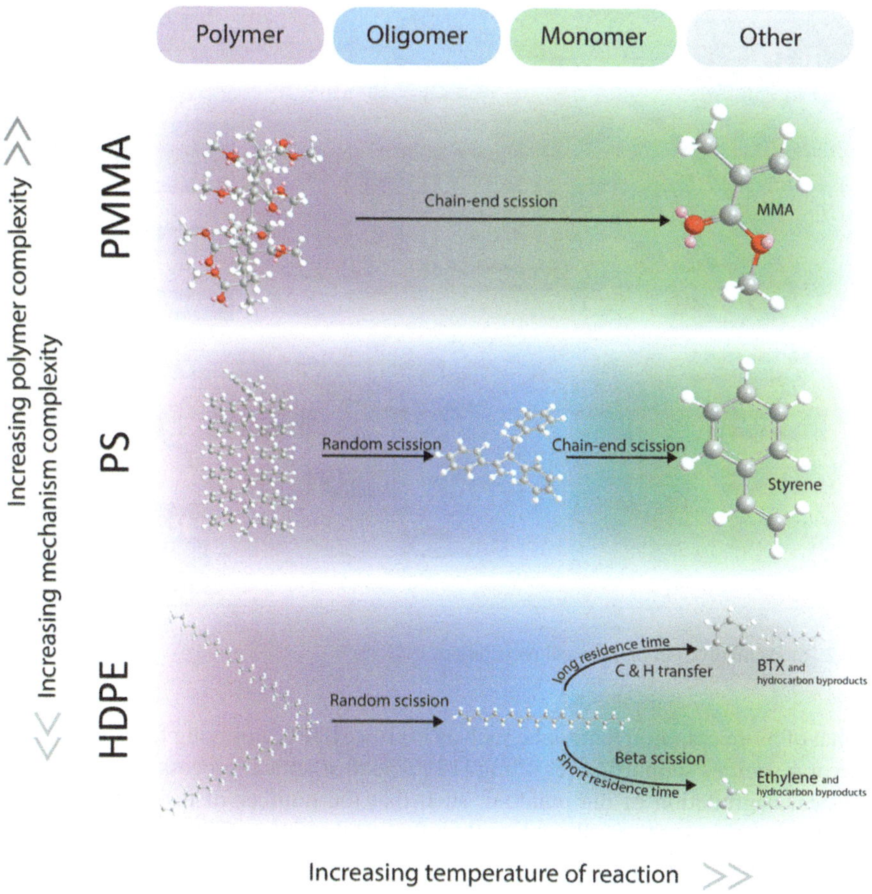

Fig. 1 Comparative analysis of degradation complexity of expanded polystyrene (EPS) with other plastic waste materials. Reproduce from an open access source [11]

3 Current Technologies for EPS Recycling

3.1 Mechanical Recycling

Mechanical recycling of Expanded Polystyrene (EPS) is a very important process in that EPS waste can be recovered and reused without any chemical structure change. The method generally involves simple steps (Fig. 2) including collection, sorting, grinding, and remolding of EPS. The mechanical recycling has some drawbacks (Table 2) also. Contamination is an issue since EPS can be commingled with other materials, which complicates the recycling process and likely degrades the quality of recycled products. Additionally, the recycling rate of EPS is relatively low compared

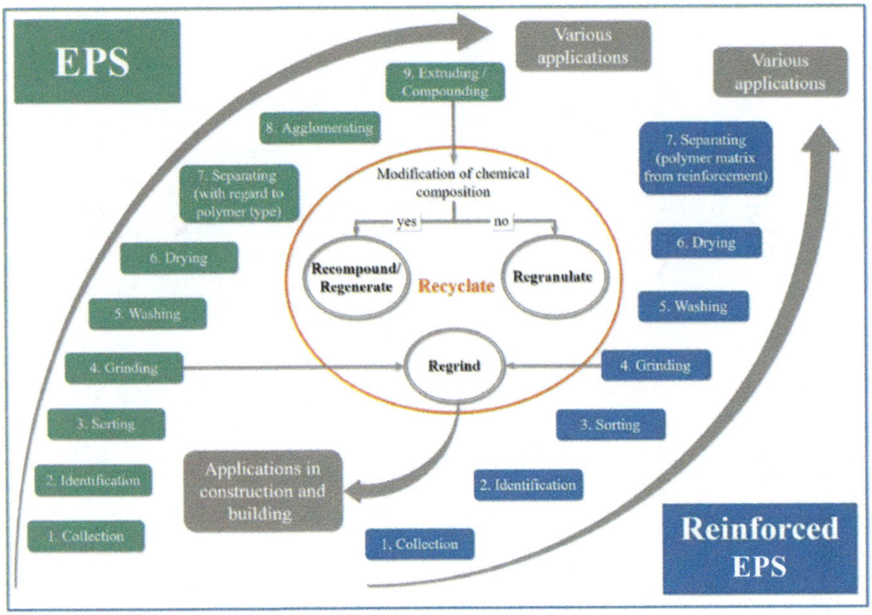

Fig. 2 Steps involved in mechanical recycling of EPS

to that of more common materials, such as PET or HDPE, implying fewer infrastructure or resources supporting recycling [12]. The mechanical process may also cause physical degradation of the material, such that the number of times that material may be recycled successfully is less than the ideal number. Therefore, mechanical recycling of EPS is a promising route for waste management and conservation of resources, but several issues face it to better proficiency and effectiveness.

3.2 *Chemical Recycling*

This involves breaking down these polymers, like EPS, into their constituent monomers or other useful chemicals through various chemical processes. It is also different from mechanical recycling due to the process used during such recycling (Fig. 3). While, chemical recycling of EPS has many benefits (Table 3). First, it saves the resources because waste EPS is converted back into valuable monomers and chemicals instead of relying on pristine fossil fuels in the production of plastics. The process supports the circular economy, where materials are utilized multiple times instead of being disposed and subsequently promoting less waste and environmental degradation. This indicates that chemical recycling also saves the material quality since the monomers attained are very pure, thus suitable for remanufacturing new EPS or any other high-grade plastics. Furthermore, the kinetics of polystyrene (PS)

Table 2 Mechanical recycling of expanded polystyrene (EPS)

Aspect	Description	Scientific outcomes	Economic impact	Ref.
Process overview	Mechanical recycling involves the physical breakdown of EPS into smaller particles or flakes, which are then cleaned and remolded into new products. This method preserves the polymer structure to a significant extent	– Average recycling yield of 85% of processed EPS – Thermal conductivity retained at 0.030–0.040 W/m·K	Estimated market value of recycled EPS projected to reach $1 billion by 2026	[13]
Key steps	1. Collection of EPS waste is done from packaging and construction sectors 2. Manual or automated sorting to separate EPS from contaminants (e.g., food residue, adhesives) 3. Waste EPS is mechanically shredded into flakes (size typically <25 mm)	– Compressive strength post-recycling typically ranges from 200 to 400 kPa – Over 80% retention of original material properties	Reduction in raw material costs for manufacturers utilizing recycled EPS, potentially lowering product prices by 15–20%	[14]
Advantages	– Average cost of mechanical recycling is approximately $0.20/kg, significantly lower than chemical recycling ($0.50–$1.00/kg) – Over 1 million tons of EPS recycled annually in the U.S. alone, significantly reducing landfill contributions – Retains over 80% of the mechanical properties of pristine EPS, making it suitable for various applications	– Mechanical recycling can maintain a tensile strength of 400–600 kPa – Life cycle assessments (LCAs) indicate a reduction of CO_2 emissions by 60%	Creation of new jobs in recycling facilities, contributing to local economies and workforce development	[15]
Challenges	– Degradation can result in a loss of impact resistance by 10–15% – Presence of contaminants can lead to product failure rates of up to 30%	– A study showed that recycled EPS experiences a 10% decrease in impact resistance compared to pristine material	Increased operational costs due to the need for advanced sorting and cleaning technologies	[16]

chemical degradation involves significant changes in its weight-average molecular weight (Mw) over time, particularly under conditions like ultraviolet (UV) radiation, thermal treatment at ambient temperature, and at 80 °C. As degradation progresses, Mw typically decreases, indicating the breakdown of polymer chains. This model elucidates the randomness of chain scission events, enhancing the understanding of PS degradation dynamics (Fig. 4). Additionally, chemical recycling can be applied

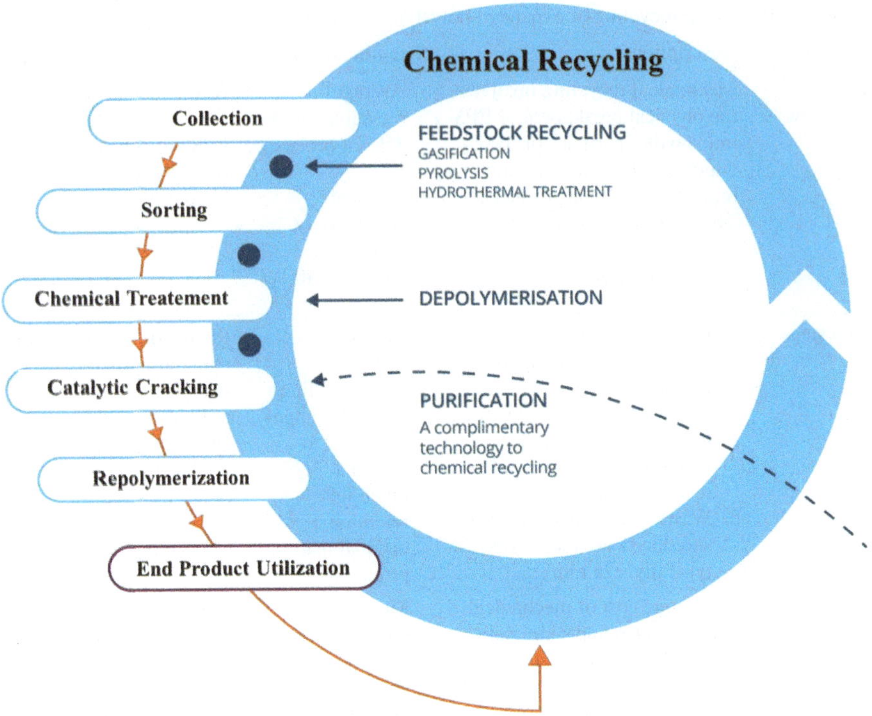

Fig. 3 Chemical recycling of EPS

effectively to various products of EPS, including those contaminated with other materials, which are unsuitable for mechanical recycling [17]. However, some challenges characterize EPS chemical recycling, which must be addressed to improve its uptake. Another factor is economic viability, chemical recycling processes are relatively very costly in terms of energy, catalysts, and specialized equipment in comparison to traditional methods of recycling [18]. Therefore, further research is needed to develop cheaper and consequently more acceptable chemical recycling technologies. As the demand for sustainable solutions grows, chemical recycling will play an increasingly important role in reducing the environmental impact of plastic waste while enabling the recovery of valuable resources.

3.3 Thermal Recycling

The thermal recycling of EPS involves breaking down the material using heat, allowing for the recovery of valuable resources and a significant reduction in landfill waste. There are several major steps involved in the process: collection and preprocessing of EPS wastes; gathering EPS from different sources, separating it from

Table 3 Chemical recycling of expanded polystyrene (EPS)

Aspect	Description	Scientific outcomes	Economic impact	Ref.
Process overview	Chemical recycling involves breaking down EPS into its chemical components, typically through pyrolysis or depolymerization, allowing for the recovery of styrene monomer or other chemicals for reuse	– High purity of recovered styrene can exceed 90% – Yield rates from pyrolysis can reach 80%	Market for recycled styrene projected to reach $2 billion by 2025, driven by demand for sustainable materials	[19]
Key steps	1. Feedstock Preparation: EPS is sorted and cleaned to remove contaminants 2. Size Reduction: Shredding the EPS into smaller pieces to enhance chemical accessibility 3. Depolymerization: EPS is subjected to high temperatures (350–800 °C) in an inert atmosphere to break down into styrene and other byproducts	– Styrene recovery rates can reach up to 95% with proper process optimization – Generated byproducts can also be utilized, providing additional economic benefits	Increased operational costs due to the need for high-temperature equipment and sophisticated purification processes	[20]
Material properties post-recycling	– Recovered styrene can be polymerized back into high-quality EPS with properties comparable to pristine material – Additional chemical byproducts can include aromatic hydrocarbons, which have various industrial applications	– Mechanical properties of recycled EPS can match those of pristine EPS, facilitating its use in high-performance applications	Potential for reduced costs in production of new EPS by using recovered styrene, potentially decreasing prices by 10–15%	[21]

(continued)

Table 3 (continued)

Aspect	Description	Scientific outcomes	Economic impact	Ref.
Advantages	– Greater than 90% recovery of raw materials compared to mechanical recycling – Can handle contaminated EPS that may not be suitable for mechanical recycling	– Studies show that chemical recycling can achieve >90% purity in recovered styrene, suitable for repolymerization	Market potential for chemical recycling technologies is expected to grow significantly	[22]
Challenges	– The need for significant heat energy makes the process costly – Requires advanced technology and expertise, making initial investment high – Residual contaminants can affect the quality of recovered materials	– Energy consumption can be as high as 2.5 GJ/ton of EPS processed, affecting economic viability	Need for economic incentives to encourage investment in chemical recycling facilities and technology	[23]

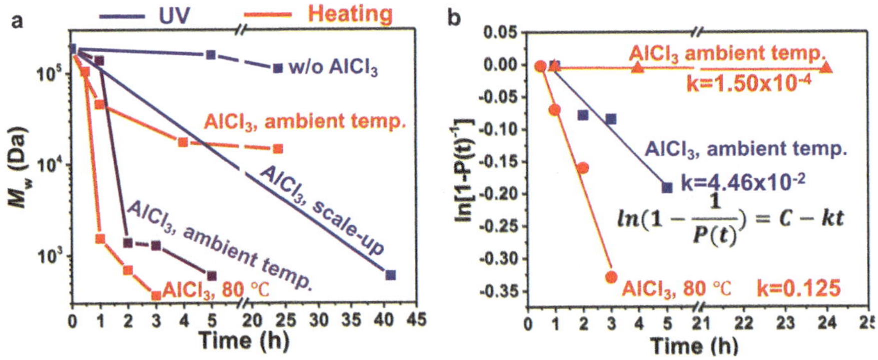

Fig. 4 Kinetics of PS chemical degradation **a** changes in the weight-average molecular weight (Mw) of PS during degradation **b** the temporal changes in Mw are modeled according to the random scission kinetic model. Reproduce from an open access source [24]

contaminants, mechanically shredding the EPS material into smaller particles to facilitate heating, controlled thermal decomposition of the preprocessed shredded EPS (Fig. 5). Normal heating is usually done at 250–400 °C where the polymer undergoes thermal depolymerization and breaks down into its monomers primarily styrene along with other products. This reaction may be carried out in a vacuum or an inert atmosphere to enhance yields and reduce oxidation of desired products. Volatile constituents are condensed during the degradation (Fig. 6). In addition, monomers styrene, that could have been used again for making new EPS can be separated and recovered and recycled after purification [25]. Residual residues remaining from this process have applications in filler contents within construction materials and composite formulations and thereby contributing toward better resource recovery and reduced wastes [26].

High energy consumption during the thermal degradation process is one of the major concerns (Table 4). Process energy efficiency is critical to the feasibility of the process in terms of economics. There must be an efficient supply chain and a market for the products to ensure commercial feasibility. Ways of increasing recycling rates include investment in infrastructure, such as recycling facilities and advanced technologies for improved efficiency in the thermal recycling process.

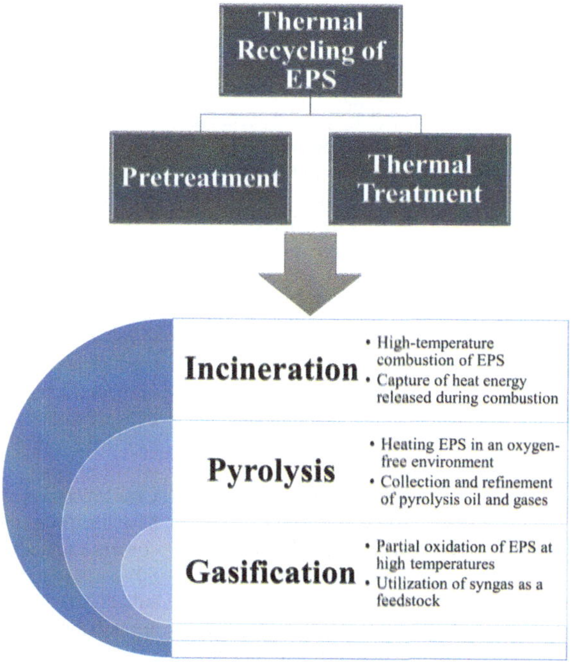

Fig. 5 Steps involved in thermal recycling of EPS

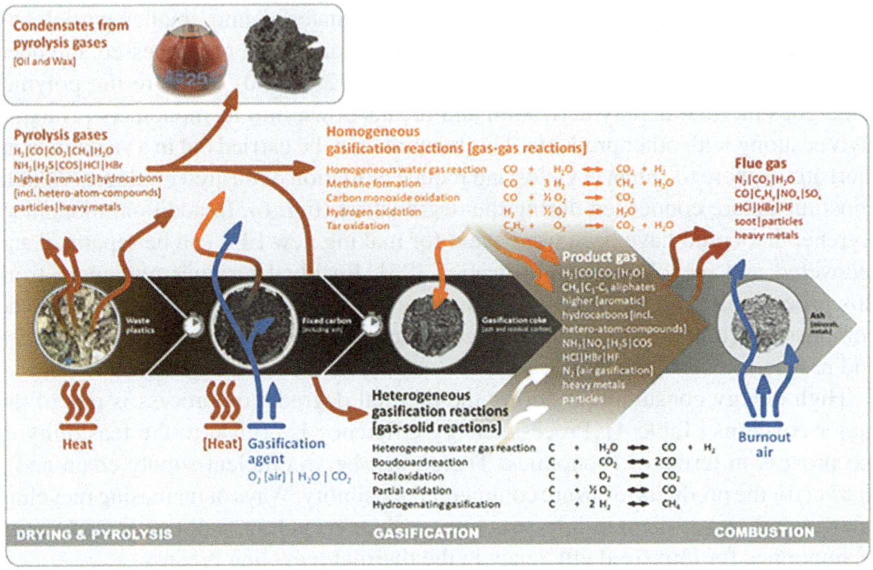

Fig. 6 Thermal recycling of EPS with evolution of gaseous components. Reproduce from an open access source [27]

Table 4 Thermal recycling of expanded polystyrene (EPS)

Aspect	Description	Scientific outcomes	Economic impact	Ref.
Process overview	Thermal recycling involves the pyrolysis or gasification of EPS to convert it into fuels, gases, or char, effectively recovering energy from waste EPS	– Typical energy recovery rate ranges from 70 to 90%, depending on process efficiency	The global market for recovered energy from plastic waste is expected to reach \$3 billion by 2025	[28]
Key steps	1. Pyrolysis: EPS is subjected to high temperatures (350–600 °C) in an oxygen-limited environment 2. Condensation: Vapors are condensed to recover oils, while gases can be utilized for energy 3. Char Recovery: Solid residues (char) can be processed further or used as a carbon source	– Pyrolysis of EPS can produce around 1 kg of oil per 2.5 kg of EPS processed, with a calorific value of 40–45 MJ/kg – Char produced can have a carbon content exceeding 80%, making it suitable for various applications	Reduced dependency on fossil fuels through energy recovery, with potential savings of \$0.15–\$0.30 per kg of EPS processed	[29]

(continued)

Table 4 (continued)

Aspect	Description	Scientific outcomes	Economic impact	Ref.
Advantages	– High Energy Recovery: Efficient conversion of EPS into valuable fuels and chemicals – Waste Volume Reduction: Up to 90% reduction in waste volume, minimizing landfill use	– Studies indicate that thermal recycling can lead to a 50% reduction in overall greenhouse gas emissions compared to landfill disposal	Potential for generating revenue from energy sales, contributing to the profitability of recycling facilities	[30]
Challenges	– Emissions Control: Requires advanced emission control technologies to mitigate pollutants (e.g., dioxins, furans) – High Energy Input: Thermal processes often require significant energy inputs, which can impact overall efficiency – Feedstock Contamination: Presence of additives or contaminants can complicate the thermal recycling process	– Emission levels can vary significantly; optimized processes can achieve up to 95% reduction in harmful emissions – Energy input can be as high as 20–30% of the total output, requiring careful process design	Initial capital investment for thermal recycling systems can be high, but long-term operational savings can offset costs	[31]

3.4 Innovative Methods in Development

Some of the innovative methods in the development of recycled EPS include Biological Recycling, Design for Disassembly (DfD), incorporation of foaming agents during EPS synthesis, Microwave-Assisted Recycling, and advanced optical sorting with IoT collection technologies. Biological recycling is one of the best solutions to handle the issues related to waste EPS. This new, innovative process of recycling focuses on the biological capabilities of microorganisms and enzymes for the degrading of EPS into simpler, degradable components. In recent years, a number of such bacteria were identified, specifically Pseudomonas putida that can utilize polystyrene as the carbon source and thus further degrade it. Other fungi, Aspergillus and Pencillium species have also been reported to break down EPS with the generation of non-toxic byproducts. Enzymatic mechanisms are also under study as a potential mechanism of EPS degradation. Researchers are isolating and engineering specific enzymes to degrade the polymer chains of polystyrene. Polystyrene hydrolase was recently identified as one such enzyme that could be used for the degradation of EPS [32]. In this context, biocatalysts applied in this situation would help in the breakdown of EPS at more mild conditions but reduce the overall environmental load in comparison with conventional chemical recycling techniques. In this way, it may turn EPS recycling into a more sustainable and efficient method if biological mechanisms are involved.

Yet another new approach in EPS recycling involves DfD principles. Through designing the product at its final stage for easy, non-destructive disassembly, DfD assures a more efficient recycling process. In other words, manufacturers can design the product with modules that could be separated without using specific tools for disassembly and thereby enhance recyclability. This can be in the form of snap-fit or interlocking designs that can easily be taken apart, and hence there is responsible disposal. Further, DfD promotes the usage of materials which may be used together with EPS and separated easily. Adding biodegradable plastics or natural fibers to the EPS products can enhance the recycling effectiveness and avoid contamination (Fig. 7). A proactive approach toward designing EPS products helps to stretch the product lifecycle far beyond mere functionality and matches more universal objectives of sustainability. The EPS's physical properties may also be modified to increase its recyclability and functionality. The method used in the production of EPS involves using environmentally friendly foaming agents. Such agents aid in the easier breakdown in the recycling processes with better recovery rates for the polystyrene. Lastly, by combining EPS with natural fibers, producers might produce composites retaining good mechanical properties but recycled more easily. These are fewer wastes but probably contribute to the development of sustainable building materials and solutions in packaging [33]. Furthermore, researches on the physical properties of EPS have their past, which made composite materials combine EPS with other polymers [34]. They can be designed to have the desired structural properties but are more recyclable. Building from material compatibility and lifecycle considerations, manufacturers then innovate new products which may also align with sustainability principles.

One new approach in this area is microwave-assisted recycling using microwave energy for degradation of EPS (Table 5). Microwave energy accelerates the internal heating inside the EPS and better than any thermal degradation in order to degrade. As a result, it attains efficiency and rapid processing by heating only at specific regions of localized regions. Accordingly, it greatly lowers costs associated with consumption and running the process. This has been one of the best advantages of microwave-assisted recycling: the contamination during the recycling process is reduced. Selective targeting of EPS means that additives or coatings are not degraded in the process, providing better quality products recycled from it. This new approach not only promotes the efficiency of recycling but also provides new means to integrate EPS into more sustainable manufacturing processes [36].

The collection and sorting technologies could impact recycling rates of EPS highly. Automated sorting systems based on AI and machine learning algorithms are increasingly being implemented in the recycling facilities. Because it makes use of physical features, it will be possible to identify and separate EPS from mixed streams, thereby allowing the recycling efficiency to increase. This involves optical

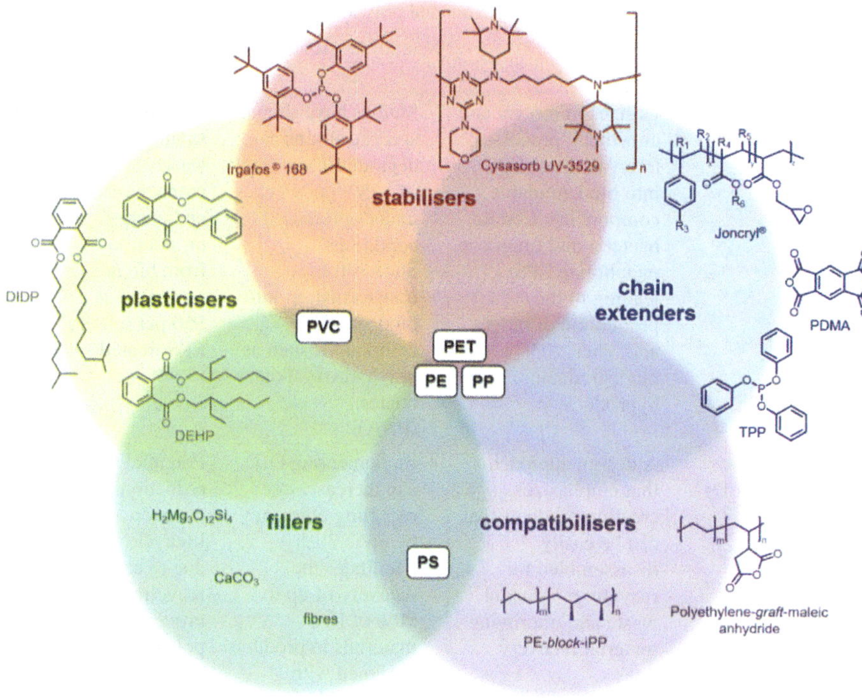

Fig. 7 Typical polymer additives utilized to enhance the properties of recycled EPS. Reproduce from an open access source [35]

sorters that increase the EPS detection and separation precision based on infrared spectroscopy. Sensors and IoT technology could also be incorporated with smart collection bins to make it more efficient to collect the EPS. Such containers monitor fill levels and rates of contamination, which assists in applying data-driven waste management techniques. These will help the municipalities optimize their routes and schedules for the collection of EPS and collecting it in time before contamination or overflows occur. Such developments in collection and sorting technologies are important aspects in upgrading the overall effectiveness and efficiency of EPS recycling efforts.

4 Applications of Recycled EPS in Packaging

Recycled Expanded Polystyrene (EPS) is gaining value due to its versatility in packaging applications because of its lightweight structure, mechanical properties, and durability. EPS is very low in density. The cost of transport and energy consumed will be quite less. The structure has many air pockets that make the EPS have unique

Table 5 Innovative methods for recycling expanded polystyrene (EPS)

Aspect	Description	Scientific outcomes	Economic impact	Ref.
Biological recycling	Utilizes microbial and enzymatic processes to break down EPS into biodegradable components. Certain bacteria and fungi can metabolize EPS, leading to the production of biogas and other biodegradable byproducts	– Studies have shown specific strains can degrade EPS at rates of 1–5% per week under optimal conditions – Successful transformation into biodegradable compounds such as polyhydroxyalkanoates (PHA)	Potential cost savings in waste management and energy production from biogas, estimated at $50 per ton of EPS treated	[37]
Design for disassembly (DfD)	A design approach that emphasizes creating products that can be easily disassembled for recycling, reducing waste and improving material recovery rates	– Implementing DfD can increase recycling efficiency by up to 30% – Facilitates the recovery of up to 90% of EPS materials in products designed with disassembly in mind	Potential reduction in product lifecycle costs due to easier recycling processes, potentially lowering costs by 15%	[38]
Incorporation of eco-friendly foaming agents	Use of environmentally benign foaming agents (e.g., CO_2 or water-based systems) instead of traditional hydrofluorocarbons (HFCs) during EPS production, reducing environmental impact during manufacturing and disposal	– The use of CO_2 as a foaming agent can reduce greenhouse gas emissions during production by 30–50% – Enhanced biodegradability of the resulting EPS can increase recycling rates	Expected reduction in production costs due to lower regulatory compliance costs and potential government incentives for sustainable practices	[39]

(continued)

Table 5 (continued)

Aspect	Description	Scientific outcomes	Economic impact	Ref.
Microwave-assisted recycling	A process that uses microwave energy to enhance the breakdown of EPS into reusable materials, such as styrene, through selective heating. This method improves efficiency by significantly reducing processing times and energy consumption	– Laboratory studies indicate recovery rates can exceed 85% with reduced processing times (by up to 50%) – Enables the recycling of contaminated EPS that traditional methods may reject	Initial investment in microwave technology is offset by reduced operational costs and increased material recovery rates, leading to a projected ROI of 20%	[40]
Artificial intelligence and machine learning (AI & ML)	Utilization of AI and ML algorithms for optimizing the recycling process, enhancing sorting efficiency, and predicting material behavior during recycling. This technology can improve decision-making in recycling facilities and enhance operational efficiency	– Implementation of AI systems can increase sorting accuracy by 90% and reduce contamination levels in recyclables – Predictive models can optimize processing conditions, leading to a 25% increase in recovery rates	Cost savings from reduced labor in sorting processes and enhanced recycling efficiency, potentially lowering operational costs by up to 30%	[41]

impact resistance, which helps absorb shock and vibrations with great efficacy. It is not prone to corrosion and is resistant to environmental conditions because of moisture presence; hence, its performance will be stable in indoor or outdoor utilization. For such kinds of durability, the storage with long-term usage application can be allowed to the recycled EPS. Further, it is easy to accept for various industries, including food, electronics, or any other industrial applications for the packaging of the recycled EPS.

4.1 Food and Beverage Packaging

Recycled EPS is widely utilized in the food and beverage industry due to its exceptional insulation properties. It is also an ideal material to use when packaging perishable goods, such as seafood, meat, dairy, and frozen foods. Because EPS is so light and retains temperature well, the packaged goods arrive at their destinations at the correct temperatures. In turn, the likelihood of spoilage of such perishable products during transport will be dramatically minimized. An example would be transporting ice cream or frozen dinners by using insulated EPS containers so that it would be kept frozen all throughout shipping. The fact that recycled EPS is not toxic and does not carry the characteristic of bacterial growth gives EPS containers a kind of hygiene advantage with foods [42]. The material does not permit sogging, ensuring a retention of food product quality. EPS can take almost any shape and form into different sizes to match food products, hence providing tailored compartments that prevent items inside the packaging from moving about. This customization is particularly beneficial for delicate items like fruits and vegetables, where bruising or damage must be minimized during transport.

4.2 Electronics and Fragile Goods Packaging

In the electronics sector, recycled EPS excels due to its shock-absorbing capabilities, making it ideal for packaging sensitive and fragile items such as computers, smartphones, and glassware. Impacts and vibrations that are likely to result from handing and transit cause damage that is exactly what EPS prevents with its cushioning feature. This is pretty significant for high-value electronics, which require secure packaging if they are to arrive intact. Custom-molded EPS inserts can cradle products to prevent movement or jostling, thus ensuring that the products stay in place and protected. In addition, post-consumer EPS can be processed so that it does not have static electricity to avoid destroying electronic products. This is beneficial in packaging sensitive appliances that might be negatively charged when exposed to static discharge [43]. The low weight of post-consumer EPS also saves on freight costs-the other advantage of using this material. In e-commerce, the freight cost apart from cheap shipping materials having protection features must be saved. That industry helps companies to suit their packaging strategy to a sustainability goal that appeals toward an environmentally conscious consumer that uses recycled material.

4.3 Industrial and Commercial Packaging Solutions

Due to its heavy-duty protection, recycled EPS is used very widely in industrial and commercial packaging. It is used for packaging of bigger items that require extra stability in the course of their transport. This includes machinery, automotive parts, and construction materials, all which are more likely to get damaged through compression resistance in EPS. In addition to this, recycled EPS also provides the industries with custom options. It can be molded to create specific inserts that will secure particular shapes. For example, appliances and industrial equipment, it will ensure the best protection by preventing the items from shifting during transport and preventing any possible damage [44]. The cost effectiveness of the recycled EPS is another added advantage, it is relatively inexpensive to produce, and can easily be molded, allowing companies to package products securely without incurring a lot of costs.

5 Environmental and Economic Impact of Recycled EPS Packaging

Recycling of EPS packaging provides much environmental and economic benefits as a result of carbon emissions decrease and resource saving to support waste reduction. Cost-effectiveness and growing market in recycled materials make reusing EPS from the recyclable ones an even feasible alternative source of sustainable packaging product. Life Cycle Assessments equally play a vital role by demonstrating such benefits to take firms in a more friendly direction toward the circular economy.

5.1 Reducing Carbon Footprint and Resource Consumption

It can save a good deal of carbon emissions produced in making it. Actually, production of pristine EPS from raw material is one of the energy-intensive processes and has heavy carbon dioxide footprints. Normally, recycling requires half to three-quarters less energy than this. Recycling, for example, recycles polystyrene to produce something new takes more than 6.0 kg of CO_2, whereas it reduces in most cases less than 0.8 kg of CO_2 per kilogram to recycle. It is that huge reduction that would definitely reduce climate change and fit entirely with the world's initiatives to reduce greenhouse gasses. Moreover, recycling EPS is an essential function in preserving non-renewable resources. Because polystyrene is a by-product of waste, the use of pristine materials-the majority of which are petroleum-based-is dramatically decreased once polystyrene is recycled. This reduces the environmental footprint in terms of resource extraction but also helps toward the circular economy where material is reused rather than waste. EPS is lightweight yet bulky, and recycling such a material minimizes landfill waste volume, thereby easing the burden on waste management systems and extending the useful life of existing landfills. The diversion of EPS from landfills is important because it is not biodegradable and would last hundreds of years, contributing to pollution and degradation of the environment [45]. An efficient recycling program of EPS can help divert incredible amounts of material from the landfills thereby solving one of the greatest problems today in waste management. Some 30% of all EPS used for packaging will end up as landfill material [46]. Using an effective recycling program, waste output is going to be lowered significantly among companies, helping to create a sustainable way of dealing with packaging matters. Additionally, reducing the volume of EPS in landfills mitigates environmental concerns related to waste decomposition, which can produce harmful greenhouse gases like methane and toxic leachates.

5.2 Cost-Effectiveness and Market Demand

From an economic perspective, recycled EPS packaging presents substantial cost advantages over its pristine counterpart. Lower material and energy costs in the production of recycled EPS often enable manufacturers to offer competitive pricing. Studies indicate that, depending on various market factors and local recycling infrastructure, recycled EPS can be produced at costs 10–30% lower than those of pristine EPS. This makes it economical, and the companies are enjoying reduced disposal fees, especially at a time when landfill expenses are rising in areas owing to strict regulations or a deficiency in capacity. Increasing market demand for green packaging options further makes recycled EPS financially feasible. With consumers turning more environmentally aware, consumers are moving toward brands who care about the environment. Companies that use recycled EPS enhance their corporate social responsibility profile while differentiating themselves in a highly competitive marketplace. Besides, many other industries, such as construction, automobile, and consumer goods, are now considering using recycled EPS for several applications, which further enhances the market growth and opens up new business opportunities. Government support and regulatory frameworks are increasingly in favor of the recycling industry. In many countries, the government is offering incentives in terms of tax breaks or subsidies to those companies that have used recycled material in the product line [47]. Along with consumers' growing expectations, legislative support becomes favorable for recycling EPS. Stringent laws with regard to plastic waste are also promoting the use of a more sustainable alternative, as demanded by the companies, thereby increasing demand for recycled EPS as an available alternative.

5.3 Life Cycle Assessment of Recycled EPS Products

A comprehensive Life Cycle Assessment (LCA) is essential for understanding the environmental impacts of recycled EPS products throughout their entire lifecycle. LCA tracks all environmental impacts associated with each life stage of a product. This can start from raw extraction to the production process, during delivery, use, and up to disposal. This presents the general ecological footprint left in recycling EPS, thus giving a chance to consumers as well as manufacturers to make the right material choices. LCA's also showed that usually the environmental impact of the recycled EPS is lower compared to pristine EPS and alternatives. For example, there is a study indicating that the production of recycled EPS consumes less energy, emits fewer greenhouse gases, and extracts fewer resources compared to the case using pristine materials as well as a few other alternatives for packaging. Such a comparative analysis gives companies a yardstick against which they can measure their sustainability of packaging decisions. Similarly, LCA also becomes an important decision-making tool for stakeholders in the supply chain. As a result, manufacturers will be able to make improvements in the process, develop better designs for the products, and work toward reducing the impact on the environment [48]. It can help policymakers make supporting regulations that will align with stewardship of the environment. When regulatory pressures augment pressure on product sustainability, LCAs can help in getting ahead of the compliance curve for companies and may avoid hefty fines or even reputational damage.

6 Challenges and Future Perspectives

Technical as well as market-wise, many serious challenges face the application of recycled EPS in packaging. Quality control should form one of the other prime technical challenges. Used or recycled EPS must be superior in quality or quality compared to that of a new material in relation to market entry. So the mechanical strength, thermostability, and overall productivity of a recycled EP shall also be at an appropriate level like any pristine material. Lower, unreliable, and fluctuating quality declines the acceptance of the producer with the products. Problem like contamination is also another technical problem. The thing is that EPS are likely mixed up with all types of contaminants throughout the life cycle, starting from food waste to other plastics and many things that compromise its integrity. Methods for sorting and cleaning would be needed to improve the quality of recycled EPS for material production. Another challenge is processing limitations. Some of the traditional recycling methods involve grinding and melting. After several cycles, this may degrade the properties of the material, therefore lowering the quality of the recycled EPS. Therefore, new methods that may be developed to keep EPS from degrading while processing are needed. For example, solvent-based recycling or supercritical fluid processes may offer superior results but will need additional research and investment before it becomes more commercially viable. Beyond these technical barriers, there are also market barriers to be overcome. Higher processing costs generally increase the price of recycled EPS compared to pristine EPS. The cost penalties associated with these increases in processing costs render it difficult for recycled materials to compete on price. Recycling processes are usually costlier and will discourage the adoption of recycled EPS when pristine materials are relatively more economical. Consumer acceptance is also another huge market barrier. Many consumers resent accepting products whose material content is pristine because of what they perceive about the quality and safety of the material. There is a big need for awareness campaigns about how recycled EPS supports environmental benefits and its performance attributes. Raising consumer's awareness about the benefits associated with using recycled materials goes a long way in encouraging market demand to shift to sustainable packaging solutions. Lastly, supply chain management logistics also present some challenges. In this case, EPS happens to be very light, yet very bulky, resulting in high cost and difficult transportation. This dispersed nature of EPS waste makes it difficult to create a steady supply of recycled material. Localized recycling initiatives and improvements in collection infrastructure can help reduce the logistical issues, leading to a better supply of recycled EPS in the market.

Besides, the future of EPS usage in packaging appears very promising and even more so if applied within the context of sustainable packaging solutions. Adopting circular economy models would mean using the highest volumes of recycled material possible while preventing as much waste as is conceivable. In the case of a circular economy, packaging has been structured so that the product is aimed for re-use and recycling with a system that would recycle back EPS into production. This approach is resource efficient and also forms a means of contributing to environmental sustainability. Further research into prospects for biodegradable alternatives made along with recycled EPS could open the way for further reduction in environmental effects by novel packaging ideas. Bio-based additives or coatings could increase the biodegradable aspects of EPS packaging in ways that keep its prime functionalities, appealing to environmentally concerned consumers and businesses. Even government policies and market trends can be in favor of increasing usage of recyclable products. Governments of different countries have formulated policies to permit the rising demand for use of recyclable EPS, which results in the formation and development of infrastructure, such as tax exemptions provided by some governments to companies using products with higher percentages of recyclable materials and quotatian for recycling provided by these governments to encourage businesses to become resource-friendly and produce more recycling-friendly end products. Corporate responsibility initiatives increase, and companies set their sustainability goals, which now include the use of recycled materials in products. It not only increases brand image but also appeals to consumers that take care about environmental practices. The advancement of technology can shape the future of recycled EPS. The integration of artificial intelligence and automation into the systems of sorting and processing should improve the efficiency of operations in recycling. The implementation of AI algorithms will refine the sorting of EPS from mixed waste streams, with automation speeding up the processes involved in further steps such as processing. More critical development of LCA tools will be achievable as well since companies will consider assessments of environmental impacts at any stage of the EPS recycle lifecycle. This transparency will thereby facilitate decision-making and support the use of recycled material through proof of sustainability claims.

7 Conclusion

This chapter has outlined immense opportunities for recycled expanded polystyrene in sustainable packaging. Some environmental advantages that come through the recycling of EPS include reduction in the amount of waste directed to landfill as well as conserving resources in terms of the reuse of the same EPS material recycled, therefore reducing carbon emissions tied to the production of new material. It is insulating, light in weight, and water-resistant, protecting the contents of products throughout their transportation and storage; recyclable EPS is thus a versatile candidate. From the perspective of economics, the recycling of EPS saves cost and, based on previous evidence, creates some employment opportunities in the recycling industries and firms involved in manufacturing. Superior technologies in recycling have developed to higher qualities with enhanced material availability. Manufacturers, recycler, and policymakers together can bring effective recycling system and more utilization of recycle material in their operations. The package industry can lead toward an ecologically good future and solve many other environmental issues through increased demand coming from environmentally conscious consumers.

References

1. Kassim N et al (2023) Sustainable packaging design for molded expanded polystyrene cushion. Materials 16(4):1723. https://doi.org/10.3390/ma16041723
2. Luz MS da et al (2023) Expanded polystyrene (eps) recycling: an environmentally friendly process to recover dissolved polystyrene (PS) using waste cooking soybean oil. Holos 5(39). https://doi.org/10.15628/holos.2023.16351
3. Bumanis Get al (2023) Thermal and sound insulation properties of recycled expanded polystyrene granule and gypsum composites. Recycling 8(1):19. https://doi.org/10.3390/recycling8010019
4. Van Deursen C et al (2023) Development of polymeric binder from expanded polystyrene (EPS) foam waste as construction materials. IOP Conf Series: Earth Environ Sci 1146(1):012007. https://doi.org/10.1088/1755-1315/1146/1/012007
5. Rodrigues C, König J, Freire F (2023) Prospective life cycle assessment of a novel building system with improved foam glass incorporating high recycled content. Sustain Product Consumpt 36:161–170. https://doi.org/10.1016/j.spc.2023.01.002
6. Lu Z et al (2023) Assessment of compressive and flexural properties and stacking strength of expanded polystyrene boxes: experimental and simulation study. Appl Sci 13(10):5852. https://doi.org/10.3390/app13105852
7. Gerassimidou S et al (2023) Unpacking the complexity of the polyethylene food contact articles value chain: a chemicals perspective. J Hazardous Mater 454:131422. https://doi.org/10.1016/j.jhazmat.2023.131422
8. Li X et al (2024) From waste plastic to artificial lotus leaf: upcycling waste polypropylene to superhydrophobic spheres with hierarchical micro/nanostructure. Appl Catalysis B: Environ 342:123378. https://doi.org/10.1016/j.apcatb.2023.123378
9. Agarwal A et al (2023) Food packaging materials with special reference to biopolymers-properties and applications. Chemistry Africa 6(1):117–144. https://doi.org/10.1007/s42250-022-00446-w
10. Ahuja A, Samyn P, Rastogi VK (2024) Paper bottles: potential to replace conventional packaging for liquid products. Biomass Conversion and Biorefinery 14(13):13779–13805. https://doi.org/10.1007/s13399-022-03642-3
11. Vollmer I et al (2020) Beyond mechanical recycling: giving new life to plastic waste. Angewandte Chemie International Edition 59(36):15402–15423. https://doi.org/10.1002/anie.201915651

12. Goshayeshi B et al (2023) Retracted: prospects of physical, mechanical, and advanced recycling of end-of-life polystyrenes: towards full recyclability. Curr Opinion Green Sustain Chem 42:100821. https://doi.org/10.1016/j.cogsc.2023.100821
13. Pardeshi S et al (2023) Mechanical recycling scenarios in India through the lens of plastic circular economy. J Mater Cycles Waste Manage 25(6):3425–3439. https://doi.org/10.1007/s10163-023-01768-8
14. Schleier J, Walther G (2024) Strategic network design for recycling of EPS insulation material—insights from a German case study. Int J Product Res 62(16):5878–5904. https://doi.org/10.1080/00207543.2024.2302041
15. Schade A et al (2024) Plastic waste recycling—a chemical recycling perspective. ACS Sustain Chem Eng 12(33):12270–12288. https://doi.org/10.1021/acssuschemeng.4c02551
16. Handtke S et al (2024) Investigation of recycled expanded polyamide beads through artificial ageing and mechanical recycling as a proof of concept for circular economy. Polymers 16(12):1730. https://doi.org/10.3390/polym16121730
17. Du J et al (2023) Complete recycling of valuable metals from electroplating sludge: Green and selective recovery of chromium. Chem Eng J 467:143484. https://doi.org/10.1016/j.cej.2023.143484
18. Celikten S (2023) Recycling waste expanded polystyrene as aggregate in production of lightweight screed mortar. Revista de la construcción 22(3):581–596. https://doi.org/10.7764/RDLC.22.3.581
19. Gu S, Xu S et al (2023) Phosphonate-influenced Diels-Alder chemistry toward multi-path recyclable, fire safe thermoset and its carbon fiber composites. EcoMat 5(9). https://doi.org/10.1002/eom2.12388
20. Xayachak T et al (2023) White spill: life cycle assessment approach to managing marine EPS litter from flood-released pontoons. Chemosphere 337:139400. https://doi.org/10.1016/j.chemosphere.2023.139400
21. Poiré-De la Cruz DR et al (2023) Flow variation impact in the synthesis of carbon nanomaterials from polystyrene recycling. MRS Advances 8(23):1315–1318. https://doi.org/10.1557/s43580-023-00672-5
22. Gu S, Xiao Y et al (2023) Neighboring molecular engineering in diels–alder chemistry enabling easily recyclable carbon fiber reinforced composites. Angewandte Chemie 135(51). https://doi.org/10.1002/ange.202312638
23. Wang C et al (2023) Converting waste expanded polystyrene into higher-value-added hypercrosslinked porous polymer for rapid and high-efficient adsorption of aflatoxins. J Clean Product 408:137102. https://doi.org/10.1016/j.jclepro.2023.137102
24. Xu Z et al (2024) Progress and challenges in polystyrene recycling and upcycling. ChemSusChem. https://doi.org/10.1002/cssc.202400474
25. Hu T et al (2024) To what content the chemical characteristics of sludge extracellular polymeric substances affect anaerobic digestion efficiency: a review. Chem Eng J 498:155490. https://doi.org/10.1016/j.cej.2024.155490
26. Acevedo-Sánchez CD, Villaquirán-Caicedo MA, Marmolejo-Rebellón LF (2023) Recycling of eps foam and demolition wastes in the preparation of ecofriendly render mortars with thermal-acoustic insulation properties. Materiales de Construcción 73(351):e317. https://doi.org/10.3989/mc.2023.342422
27. Biessey P et al (2023) Plastic waste utilization via chemical recycling: approaches, limitations, and the challenges ahead. Chemie Ingenieur Technik 95(8):1199–1214. https://doi.org/10.1002/cite.202300042
28. Foti D et al (2023) Heat treatment's effect on properties of polystyrene from building demolitions. BioResources 18(3):5623–5634. https://doi.org/10.15376/biores.18.3.5623-5634
29. Masri T et al (2023) Characterization of a composite material composed by rubber tire and expanded polystyrene wastes. Revue des composites et des matériaux avancés 33(1):13–19. https://doi.org/10.18280/rcma.330103
30. El Gamal S et al (2024) Mechanical and thermal properties of lightweight concrete with recycled expanded polystyrene beads. European J Environ Civil Eng 28(1):80–94. https://doi.org/10.1080/19648189.2023.2200830

31. Biçer A (2023) The thermal analysis of plaster with waste expanded polystyrene gypsum and resin. Int J Innovative Eng Appl 7(1):113–118. https://doi.org/10.46460/ijiea.1212678
32. Zhong W et al (2024) PS/PPO foam with excellent comprehensive properties prepared by integrated foaming and sintering strategy with microwave-assisted. Composites Communications 48:101925. https://doi.org/10.1016/j.coco.2024.101925
33. Shojaei S, Rostami-Tapeh-Esmaeil E, Rodrigue D (2023) Polymer foams waste management: a focus on mechanical and chemical recycling 289–318. https://doi.org/10.1021/bk-2023-1440.ch014
34. Yu M-Y et al (2024) Effect of microwaves combined with peracetic acid to improve the dewatering performance of residual sludge. Environ Sci Pollut Res 31(32):44885–44899. https://doi.org/10.1007/s11356-024-33931-5
35. Schyns ZOG, Shaver MP (2021) Mechanical recycling of packaging plastics: a review. Macromol Rapid Commun 42(3). https://doi.org/10.1002/marc.202000415
36. Yang C et al (2023) A review on the microwave-assisted pyrolysis of waste plastics. Processes 11(5):1487. https://doi.org/10.3390/pr11051487
37. Jiang Y et al (2024) Bio-based hyperbranched epoxy resins: synthesis and recycling. Chem Soc Rev 53(2):624–655. https://doi.org/10.1039/D3CS00713H
38. Li X et al (2023) Volatile fatty acids bio-production using extracellular polymeric substances disengaged from sludge for carbon source recycling. Bioresour Technol 386:129565. https://doi.org/10.1016/j.biortech.2023.129565
39. Bharathi SD et al (2023) Resource recycling, recovery, and xenobiotic remediation from e-wastes through biofilm technology: a review. Appl Biochem Biotechnol 195(9):5669–5692. https://doi.org/10.1007/s12010-022-04055-8
40. Franco-Morgado M et al (2023) Microalgae and cyanobacteria polysaccharides: important link for nutrient recycling and revalorization of agro-industrial wastewater. Appl Food Res 3(1):100296. https://doi.org/10.1016/j.afres.2023.100296
41. Rybak J, Stojanowska A, Zeynalli F (2023) Biodegradability and bioremediation of polystyrene-based pollutants. In: Biodegradability of conventional plastics. Elsevier, pp 179–200. https://doi.org/10.1016/B978-0-323-89858-4.00004-X
42. Daugėlaitė V, Kruopienė J (2024) Barriers of and possibilities for recycling of single-use take-away food and beverage packaging: evidence from lithuanian market. Sustainability 16(13):5732. https://doi.org/10.3390/su16135732
43. Kim SY et al (2023) Comparative life cycle assessment of reusable and disposable distribution packaging for fresh food. Sustainability 15(23):16448. https://doi.org/10.3390/su152316448
44. Chan RBY (2023) Drivers of divergent industry and consumer food waste behaviors: the case of reclosable and resealable packaging. J Clean Prod 412:137417. https://doi.org/10.1016/j.jclepro.2023.137417
45. Young E, Mirosa M, Bremer P (2023) A conceptual model for food industry views on the commercialisation of active and intelligent packaging. Packag Technol Sci 36(11):905–925. https://doi.org/10.1002/pts.2766
46. Zaragoza-Benzal A et al (2024) Upcycling EPS waste and mineral wool to produce new lightweight gypsum composites with improved thermal performance. Construct Build Mater 449:138464. https://doi.org/10.1016/j.conbuildmat.2024.138464
47. Trieu K et al (2024) Estimated health benefits, costs, and cost-effectiveness of implementing WHO's sodium benchmarks for packaged foods in India: a modelling study. The Lancet Public Health 9(11):e852–e860. https://doi.org/10.1016/S2468-2667(24)00221-4
48. Meng T et al (2023) Comparison of the environmental impact of typical packaging systems for food cold chain express based on life cycle assessment. J Clean Product 430:139756. https://doi.org/10.1016/j.jclepro.2023.139756

Chemical Recycling of Expanded Polystyrene Foams

Jorge Lopez-Tellez, Karen A. Escamilla, Israel S. Ibarra, and Jose A. Rodriguez

Abstract Chemical recycling of expanded polystyrene foams is tertiary recycling and considered an alternative for low-cost waste treatment because of it is simple and environmentally friendly process. Chemical recycling of expanded polystyrene is described by three routes: (1) solvent-based polystyrene dissolution (2) polystyrene structural modification and (3) polystyrene depolymerization. These strategies represent a useful option to generate new materials such as: fiber, membrane, cationic exchanger resin (flocculant materials), polyelectrolyte, and fuel cell membranes and they was explored to obtain high value aromatic products: styrene (monomers and oligomers), toluene, propenyl benzene, ethyl benzene, benzoic acid and other aromatic compounds. This chapter presented and discussed current trends, advances and alternatives for chemical recycling of polystyrene wastes.

Keywords Recycling · Expanded polystyrene · Depolymerization · Polymer modification · Solvent-based recycling

1 Introduction

Chemical recycling for polymeric materials is considered tertiary recycling [1] and has generally been described as the conversion of polymers into monomers or partially depolymerized to oligomers through a chemical reaction (depolymerization). Monomers obtained from depolymerization can be used for new polymerizations to reproduce the original or a related polymeric product [2, 3].

Chemical recycling can be classified considering the process and the type of the agents employed [4–6]. Chemolysis or solvolysis (solvent-based recycling) is based on polymers decomposition by chemical reagents to produce monomers and oligomers, reaction products can be repolymerized into new plastics. Depending

J. Lopez-Tellez · K. A. Escamilla · I. S. Ibarra · J. A. Rodriguez (✉)
Area Academica de Quimica, Universidad Autonoma del Estado de Hidalgo, Pachuca de Soto, Hidalgo, Mexico
e-mail: josear@uaeh.edu.mx

A. Khan et al. (eds.), *Uses and Products of Recycled Expanded Polystyrene Foam Wastes*, Sustainable Materials and Technology,
https://doi.org/10.1007/978-981-96-5618-9_4

on the type of chemical agent involved, the main chemolysis types for depolymerization are alcoholysis, aminolysis, hydrolysis and glycolysis [6], in each case, the reagent compounds break of hydrolysable polymer bonds, offering several possibilities depending on reaction conditions, product selectivity, yield, solvents, catalysts, temperature and pressure used [7].

Thermo-chemical recycling plastic wastes is another type of chemical recycling type in which the polymer is heated either in absence of oxygen (pyrolysis and hydrocracking) or with limited oxygen (gasification). Pyrolysis process consists in submit plastic wastes at temperature from 350.0 to 700.0 °C [8], while hydrocracking or catalytic cracking conditions involves a hydrogen atmosphere, high pressure, temperatures >350.0 °C in presence of a catalyst [4]. Gasification is the partial oxidation in the presence of steam, oxygen and air at temperatures in the interval of 800.0–1600.0 °C and pressures of 15.0–30.0 MPa, this process can produce synthesis gas or syngas [9]. Thermo-chemical process is included as thermal decomposition methods because of they depend more on the temperature, in addition, pyrolysis and catalytic cracking are techniques used in the production of fuels and oils from polymeric materials [10–12].

On the other hand, the EU Waste Framework Directive defines chemical recycling applied to polystyrene (PS) as any operation where waste materials are reprocessed into products, materials or substances, these do not include obtaining energy or reprocessing materials that will be used as fuels or backfilling operations [6, 13]. Under this context chemical recycling for PS is described in the literature by three routes (Fig. 1): (1) solvent-based PS dissolution can be recycled semi-chemical method [14], even though there is no modification in the chemical structure, recovery of the initial polymeric matrix allows the obtaining of materials that can be applied within various chemistry areas [15], (2) PS structural modification is the second route, in this treatment the PS is dissolved and the chemical structure is modified [16], and (3) depolymerization, a common recycling approach for PS through a polymer degradation to produce high value styrene monomers, oligomers and other organic compounds [17]. Notwithstanding, depolymerization occurs in pyrolysis, catalytic cracking and gasification and several authors in the literature have described them as chemical recycling, these processes do not fall within the definition according to EU Waste Framework Directive due to in the depolymerization process is produced fuel and oil, consequently they will not be included in the chemical recycling classification for polystyrene. In addition, pyrolysis, and gasification are classified as thermal decomposition as they depend on the temperature. Therefore, chemical recycling for PS strategies would be discussed in the next sections.

2 Semi-Chemical PS Recycling

First chemical PS recycling route is waste solubilization by solvents, in this case, PS does not suffer a depolymerization or rupture PS [14]. Therefore, this strategy has been defined as semi-chemical recycling, which allows a change in the PS shape and

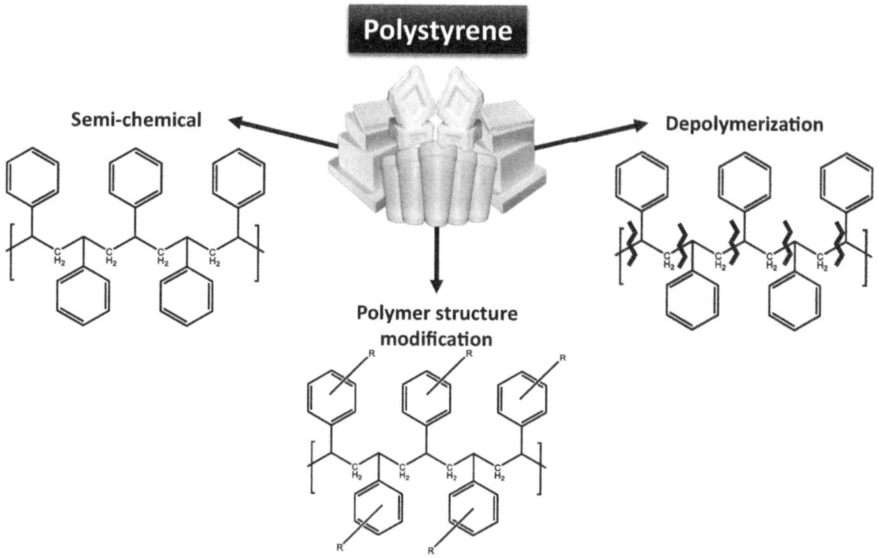

Fig. 1 Chemical recycling for PS according to EU waste framework directive definition

volume. However, PS does not present chemical structural modifications [14, 16]. PS waste solubilization does not require additional energy for material transformation, neither application of mechanical techniques, which is an advantage in reducing transportation and disposal costs. Furthermore, solubilization is a functional tool to remove adhered impurities from PS based on low solubility.

Most important step to carry out the PS dissolution is the selection of the appropriate solvent [14, 18]. Due to the similarity in physical and chemical properties, PS dissolves easily in non-polar or intermediate polarity solvents such as: xylene, toluene, benzene, nitrobenzene, cyclohexane, chloroform, tetrahydrofuran, acetone and ethyl acetate [19, 20]. The presence of aromatic ring in the solvent chemical structure allows a better PS-solvent interaction. The resulting mixture (solvent-PS) has been used as a precursor in the manufacture of new materials with the aim of giving it a second use and application. In recent years, the manufacture of materials such as films, fibers and filters fibers have been developed. PS structural characteristics allow it to be used as adsorbent material for extraction and analysis of aromatics compounds, mainly due to the ability to form π-π type interactions between aromatic rings present in PS and the compounds studied. This interaction mechanism favors the retention of the target compounds on materials manufactured from PS waste [21].

Fibers obtained from recycled PS is an adsorbent material which attracted great interest among researchers due to its easy preparation and potential applications. The most common technique to manufacture fibers is electrospinning. Generally, fibers are obtained in nanoscale and show excellent characteristics such as high surface area, high porosity, mechanical and chemical resistance, good chemical stability, interconnected ultrafine fibrous structure and high permeability which makes them a viable material for removal or filtration purposes [22].

PS nanofibers with diameter about of 400.0 nm, high porosity, large surface and high extraction efficiency were prepared by dissolving PS in a mixture of dimethylformamide and tetrahydrofuran (1:1). Extraction of antipsychotics (olanzapine, risperidone, paliperidone, clozapine, quetiapine, ziprasidone, and aripiprazole) in human plasma was carried out using this adsorbent and followed by a quantification using instrumental techniques such as high-performance liquid chromatography [23].

Besides to fibers, it is possible to obtain other materials from the PS solution with other morphology such as films that have been other characteristics that allow them in analytical methodologies. Ríos-Gómez et al. describe the manufacture of PS films supported on filter paper for the extraction and analysis of methadone in human urine samples. Yoghurt containers were used as a source of PS and chloroform to prepare the polymer solution, deposition of PS film is obtained by dipping pieces of filter paper into the solution mentioned above. PS films have provided good results in drug analysis in biological samples; by placing the material on a pipette tip, handling and sample volume required for testing are minimized [24]. Furthermore, the manufacture of this material is a simple, fast and low-cost (green chemistry). PS films have been employed in analysis of small sample volumes, such as biological samples; however, is possible working with large sample volumes reducing the risk of cross-contamination. PS from commercial containers was dissolved in tetrahydrofuran to fabricate films for polycyclic aromatic hydrocarbons analysis in water samples. The film was placed inside the tip of a pipette and 12 polycyclic aromatic hydrocarbons adsorbed and then determined by high-performance liquid chromatography [21].

Depending on the hydrophobic character of the materials manufactured from PS, other relevant application is the elimination of organic compounds present in aqueous media (dispersed oil phase). An adsorbent material was manufacture based on PS from food containers and natural cotton fibers. The procedure to manufacture this type of materials is quick and simple; an immersion of cotton fibers in a PS solution (chloroform). Through this method, the deposition of a polymer layer on the cotton fibers is favored, allowing the fiber to exhibit a high hydrophobic character. Strong interactions between fibers and aromatic compounds (fluorene or pyrene) have been described, in addition to a high separation of oil phase from aqueous media [25]. The development of adsorbent materials from natural compounds and polymeric waste turns out to be an ecological alternative due to a simple and economical way of manufacturing and the good results that have been obtained.

Air filters can be obtained from the combination of layers of microfibers and nanofibers. It has been shown that by modifying the solvent used for PS recycling it is possible to obtain fibers of various sizes, which represents an advantage for industrial applications. When dimethyl formamide is used to dissolve the polymer, it is possible to obtain fibers with an average diameter of 8.9 μm, however, by adding a biodegradable solvent such as D-limonene to the mixture of solvents, it is possible to reduce the fiber size to 1.0 μm. An air filter was manufactured based on the combination of several layers of the above-mentioned fibers, the filter has a filtration efficiency of 99.4% and a quality factor of 0.099 Pa^{-1}. These filters are promising due to the great advantages they offer such as their low cost and production time (12 min) [26]. In addition to materials where functionality is based on PS structural characteristics and properties, PS waste can be used in the manufacture of other materials where the main function is to be a support for depositing other compounds. The PS waste is a good support due to its stability, inert nature, low cost and easy accessibility. In general, the supports are manufactured by immersion of PS material in solution of the compound that will be supported.

Chitosan supported on PS nanofibers has been described for removal of Pb (II) in aqueous media, nanofibers were manufactured from the dissolution PS waste in N,N-dimethylacetamide. In a second step, it is necessary to immerse the nanofibers in a chitosan acid solution. Incorporation of ^{-}OH (hydroxyl) and $^{-}NH_2$ (amino) groups to nanofibers allows a strong interaction with metal ions, conferring a high adsorption capacity (137.4 mg g^{-1}) [27]. On other hand, PS from yoghurt packaging was used as support to immobilize TiO_2, used as a photocatalyst in the degradation of erythrosine and brilliant blue. In this case, PS waste only was washed with neutral detergent and water, followed by immersion in a 2.0% (w/v) TiO_2 aqueous suspension solution at room temperature. The main advantage of this material is the reduction of chemical agents used and that exposure to high temperatures is not necessary to carry out the degradation of the dyes, under these conditions 98.0% colorants degradation is possible [28].

It has been shown that adsorbents manufacture of supports from food packaging material is an excellent alternative for water treatment based on adsorbents with the advantage of reducing PS waste environmental impact. In summary, various materials, with important applications, have been obtained from PS waste dissolution and some applications are collected in Table 1.

Organic solvents costs are attractive to industry for their applicability in extraction, cleaning, formulation, coating, and chemical production. According to European Solven Industry Group about 5 million tons of solvents are used by the European Industry per year [41]. However, some solvents present negative consequences for health such as: teratogenicity, nephrotoxicity, hepatotoxicity, neurotoxic disorders and carcinogenicity [40].

To reduce the negative impact of organic solvents in PS, it has been proposed the use of natural and biodegradable compounds. One of the first options described is the use of D-Limonene, a terpene presents in citrus peel, due to the characteristics allows that the polymer maintains a solubility similar to organic solvents mentioned above [42]. In this sense, natural compounds with similar chemical composition have been explored for the PS dissolution. Table 2 presents various natural solvents employed with their respective PS solubilities. Essential oils have high contents of terpenes and their derivatives, these have been evaluated, presenting important advantages: developing at room temperature and do not require specialized equipment. In addition, they are relatively economical and environmentally friendly. Flowers such as jasmine, orange blossom, lavender, lily, tuberose, gardenia, violet and ylang-ylang have been used to obtain essential oils with high terpene content and improve the residues PS dissolution [43]. As a perspective, it is intended to manufacture new

Table 1 PS semi-chemical recycling and applications

Solvent	Morphology	PS solution composition (% w/v)	Application	Ref.
N,N-dimethylacetamide	Nanofibers	15.0	Filter media (emulsions)	[29]
Dimethylformamide	Nanofibers	15.0	Adsorption of organic solvents (acetonitrile)	[30]
Chloroform	Magnetic PS nanocomposite	2.0	Determination of parabens (methylparaben, ethylparaben, propylparaben and butylparaben)	[31]
Chloroform	PS-cotton composite (fibers)	2.0	PAHs remotion	[25]
Dimethylformamide, tetrahydrofuran (1:1)	Nanofibers	15.0	Analysis of sedative-hypnotic (zaleplon, zolpidem and six benzodiazepines)	[32]
Dimethylformamide, tetrahydrofuran	Fibers	20.0	Remediation of oil wastewater	[33]
Dimethylformamide, tetrahydrofuran (4:6)	Nanofibers	10.0	Extraction of pesticides (diazinon, malathion, propargite, bromopropyrate, tetradifon and permethrin)	[34]
Dimethylformamide, tetrahydrofuran (8:2)	Nanofibers	15.0	Extraction of Disulfine blue	[35]
Tetrahydrofuran	Film	–	Packaging	[36]
Acetone, ethyl acetate (20:80)	Film	–	Coating material for asphalt and infrastructure applications	[37]
Chloroform	Film	3.0	Microextraction of methadone	[24]
Tetrahydrofuran	Film	15.0	PAHs determination	[21]
Tetrahydrofuran	Coating	15.0	PAHs determination	[38]
Dimethylformamide, d-limonene (3:1)	Nanofibers	20.0	Air-filter	[39]
Dimethylformamide, d-limonene	Nanofibers	25.0	Air-filter	[40]

PAHs Polycyclic aromatic hydrocarbons

Table 2 PS solubility in several natural compounds

Solvent	PS solubility g/100.0 g solvent	Ref.
α-Terpinene	130.2	[44]
γ-Terpinene d-Limonene Terpinolene	130.6 126.7 125.2	
Geranyl acetate Bornyl acetate	174.4 67.2	
α-Pinene	43.8	
1,8-Cineole	54.8	
α-Phellandrene	125	[45]
β-Phellandrene	122	
p-Cymene	212	

materials from the PS solution of using biodegradable, low-cost and environmentally friendly compounds, as well as evaluate the functionality of these materials.

3 Chemical PS Structural Modification

Second PS chemical recycling route is based on chemical polymeric modification without depolymerization [15]. There are several reports on electrophilic substitution on styrene ring as sulfonation, nitration or amination. The modified PS products have been widely used cationic exchanger resin (flocculant materials), polyelectrolyte, and fuel cell membranes [46, 47]. Sulfonation and nitration are the main strategies employed for PS modification, the initial step is reducing the PS particle size and subsequently the polymer phase is introduced into concentrated acid (sulfuric acid or nitric acid), the reaction is carried out at temperatures around 40.0 °C under stirring, the modified PS particles are separated and washed to remove excess reagent [48]. The second method consists of two steps; first the PS is dissolved using the appropriate solvent. Subsequently, concentrated acid is added, the mixture is stirring and moderately heated (25.0–90.0 °C) [46]. Both methods require some catalyst to improve the reaction [46, 49]. The main advantages of sulfonation and nitration are the increase hydrophilicity, proton conductivity and water-solubility [50].

Other sulfonation method was described by Dardeer and Toghan PS was dissolved in chloroform; then sulfuric acid was added. The mixture was stirred for approximately 8.0 h at 50.0 °C (until the reaction turns brown), using this method a sulfonation yield of 65.0% is described [51]. Sulkowski et al. compared two sulfonation methods, the first method consists of pulverizing PS (0.40–1.04 mm), this was added to a mixture of Ag_2SO_4 as catalyst and H_2SO_4 at 80.0 °C for 2.0 h. The second method evaluated was sulfonation using silica sulfuric acid, which is added to a mixture of PS in 1,2-dichloroethane, the molar ratio of PS to sulfonating agent was

1:2 for 7.0 h at 60.0 °C [49]. Both methods of PS sulfonic acid synthesis were evaluated as flocculants for the treatment of wastewater, because they exhibit properties of anionic polyelectrolytes, obtaining an ion exchange capacity similar to commercial ion exchangers. The combination of sulfonated product with silica sulfuric acid shown better results in wastewaters treatment.

Mahmoud et al., described a nitration method from residual PS, which initially PS was pulverized until fine particles were obtained followed by a washing step with distilled water to ensure the complete elimination of the adsorbed impurities, then the PS particles were placed to react in a mixture of concentrated nitric acid and sulfuric acid for 24.0 h. The modified material was used as adsorbent to remove Cd (II), Pb (II) and Hg (II) from water samples [52]. Table 3 shows the different sulfonation and nitration modifications for PS.

Amination or other functional group addition occurs in three steps, (1) PS is dissolved in an appropriate organic solvent, (2) the carboxyl group is grafted onto styrene ring by the Friedel–Crafts acylation reaction, usually the acylation reagents are acyl chlorides and Lewis's acid catalyst ($FeCl_3$ and $AlCl_3$), and (3) and subsequently the acyl group is exchanged to derivatives of carboxylic acids. These new polymeric materials have been applied to adsorption of contaminants in several samples due to good chemical stability, high surface area and low solubility. Main interaction mechanism is π-π type interactions with various organic contaminants with aromatic groups into their structure.

Pu et al. developed PS modification with dopamine and N,N″-dicyclohexylcarbodiimide. PS was dissolved in dichloromethane with succinic anhydride and $FeCl_3$ as catalyst for addition a carboxyl group through the Friedel–Crafts, after, the dopamine and N,N″-dicyclohexylcarbodiimide was added and stirred at room temperature for 12.0 h and the $FeCl_3/FeCl_2$ in NH_3 solution was added for formation of magnetic particles at 80.0 °C for 20.0 min, the new material was used for Congo red and methylene blue extraction from wastewater [58]. Other alternative is the PS modification surface with phenyl hydrazine. PS surface modification was carried out using carbon tetrachloride as dissolvent, anhydrous $AlCl_3$ and acetyl chloride for acylation (50.0 min at 60.0 °C), afterward phenylhydrazine was immobilization for 60.0 min at 50.0 °C for the hydrazone formation. PS-hydrazine was applied in the adsorption of phenol [59].

PS hypercrosslinked polymer was synthesized, PS was dissolved in dichloromethane, formaldehyde dimethyl acetal and $FeCl_3$ were added, and the mixture was allowed to reflux for 12.0 h. The Lewis acid catalyst allowed methoxymethyl groups to attach to additional phenyl rings, these transform into methylene bonds, resulting in very rigid cross-linked structures. Several aromatic building components were directly knitted together, resulting in networks with a high surface area and a high degree of microporosity. The characteristics of this material allowed CO_2 adsorption [60]. Table 4 shows the different methods for PS structure modification.

Table 3 Sulfonation and nitration modifications for chemical PS recycling

Modification	Chemicals agents	Reaction conditions	Application	Ref.
Sulfonation	PS (0.5–1 mm) Concentrated H_2SO_4	1.5 h at 60.0 °C	Remotion Pb (II) and Cd (II)	[48]
(1) Sulfonation (2) Silica sulfonation	(1) Ag_2SO_4 (catalyst) concentrated H_2SO_4 (2) PS dissolved in 1,2-dichloroethane silica sulfuric acid	(1) 80.0 °C for 2.0 h (2) molar ratio 1:2 for 7.0 h at 60.0 °C	Flocculants	[49]
Silica sulfonation	PS dissolved in 1,2-dichloroethan silica sulfuric acid	Molar ratio 1:2 for 7.0 h at 60.0 °C	Remotion Zn (II) and Cu (II)	[53]
Nitration	Concentrated HNO_3–H_2SO_4	24.0 h at room temperature	Remotion Cd (II), Pb (II) and Hg (II)	[52]
Sulfonation	PS	1.0 h at 70.0 °C	Remotion Zn (II) and Pb (II)	[54]
Sulfonation/Chelating agent	Concentrated H_2SO_4 and 0.002 M [S,S]-ethylenediamine-N,N′-disuccinic acid	40.0 min at 80.0 °C	Remotion ciprofloxacin	[55]
Sulfonation/ Pseudopolyrotaxane	PS dissolved in chloroform Concentrated H_2SO_4 γ-cyclodextrin	8.0 h at 50.0 °C	–	[51]
Sulfonation-iron oxide nanoparticles	PS particles <4.75 mm Concentrated H_2SO_4–$FeCl_2$	1.5 h at 95.0 °C	Degradation indigo carmine	[50]
Nitration	PS dissolved 3-nitrotoluene (50.0 °C) Concentrated HNO_3–H_2SO_4	Acid was added at 10.0 °C and reaction at 30.0 °C for 2.0 h	Energy storage	[56]
Nitration-amination	(1) PS dissolved in 1,2-dichloroethane Concentrated HNO_3-H_2SO_4 (2) PS-nitrated was added in HCl/absolute ethanol and Sn granules	(1) 30.0 °C for 20.0 h (2) 50.0 °C for 9.0 h	CO_2 Capture	[57]
Sulfonation	PS dissolved cyclohexane P_2O_5/H_2SO_4	1.0 h at 40.0 °C	Coagulation-Flocculation	[46]

Table 4 Chemical modifications for chemical PS recycling by Friedel–Craft reaction and acyl group is exchanged to carboxylic acids derivatives

Carboxylic acids derivatives	Chemicals agents	Reaction conditions	Application	Ref.
Hydrazone	(1) PS dissolved in carbon tetrachloride, $AlCl_3$ and acetyl chloride (2) Phenyl hydrazine	(1) 50.0 min at 60.0 °C (2) 60.0 min at 50.0 °C	Remotion phenol	[59]
Amide	(1) PS dissolved in dichloromethane, succinic anhydride and $FeCl_3$ (2) Dopamine and N,N′′-dicyclohexylcarbodiimide	(1) Room temperature for 8.0 h (2) Room temperature for 18.0 h	Remotion organic dyes	[61]
Ketone-Natural rubber	(1) PS dissolved in tetrahydrofuran Maleic anhydride Boron trifluoride etherate (2) Maleation of PS and natural rubber dissolved in THF N-dimethyl aniline and benzoyl peroxide (catalyst)	(1) Room temperature for 24.0 h (2) Room temperature	–	[62]
Ketone	PS dissolved in methylene chloride 1,2,4-benzenetricarboxylic anhydride and $AlCl_3$	60.0 °C for 10.0 h	Remotion methylene blue, Safranin T, and Malachite green	[63]
Oxidation-Polyethylenimine immobilization	(1) PS granules in chromic acetic solution (2) Polyethylenimine in deionized water Glutaraldehyde	(1) 100.0 °C for 8.0 h (2) Room temperature for 12.0 h	Remotion humic acid	[64]
Crosslinked polymer	(1) PS was dissolved in 1,2-cichloroethane, carbon tetrachloride, formaldehyde dimethyl acetal, α,α'-dichloro-p-xylene, 4,4′-bis(chloromethyl)-1,1′-bipheny and 2) $AlCl_3$ in ethanol	(1) Room temperature (2) 12.0 h at 80 °C	CO_2 capture	[65]
Hyper crosslinked polymer	(1) PS was dissolved in 1,2-cichloroethane (2) Formaldehyde dimethyl acetal and $FeCl_3$	(1) Room temperature (2) Refluxed 24.0 h	CO_2 capture	[66]
Amide	(1) PS dissolved in dichloromethane, maleic anhydride and $AlCl_3$ (2) N,N′′-diisopropylcarbodiimide and tetraethylenepentamine	(1) 5.0 h at room temperature (2) 24.0 h at room temperature	Remotion Eriochrome Black T and Congo red	[67]

(continued)

Table 4 (continued)

Carboxylic acids derivatives	Chemicals agents	Reaction conditions	Application	Ref.
Amide-Magnetic nanoparticles	(1) PS dissolved in dichloromethane, succinic anhydride and $FeCl_3$ (2) Dopamine and N,N′′-dicyclohexylcarbodiimide (3) $FeCl_3/FeCl_2$ in NH_3 solution	(1) 8.0 h at room temperature (2) 18.0 h at Room temperature	Remotion organic dyes	[58]
Hypercrosslinked polymer	(1) PS was dissolved in dichloromethane and formaldehyde dimethyl acetal and (2) $FeCl_3$	(1) 3.0 h min at room temperature (2) 12.0 h at 40.0 °C	CO_2 capture	[60]

4 Chemical PS Depolymerization

The last route for PS chemical recycling allows obtaining small molecules. Transformation by depolymerization is defined as a change in chemical structure to convert polymeric materials into monomers, oligomers, and other hydrocarbon compounds that serve as a starting point for the new materials [68]. This methodology offers great advantages to the circular economy, since it allows obtaining high-value compounds that serve as raw materials in the production of medicines, resins or fragrances.

Due to the costs represented by the drastic conditions of the techniques, in recent years strategies have been developed using mild conditions (temperature <300.0 °C) and more economical chemical agents than conventional catalysts. A non-thermal plasma-enabled based method at ambient temperature and atmospheric pressure was developed to break C–C bonds by PS hydrogenation. PS is depolymerized using a reactor equipped with a dielectric barrier discharge plasma generator in presence of H_2 flow. PS recycling under these conditions allows obtaining high commercial value hydrocarbons, CH_4, C_2H_2, C_2H_4, C_3H_6, and C_3H_8, ethylene is the main reaction product which is an important starting material for the manufacture plastics and solvents representing an important progress to industry [69].

In addition to styrene, other molecules are obtained under ambient conditions from the PS chemical depolymerization such as toluene, propenyl benzene, ethyl benzene, and other aromatic compounds. In this sense, a novel alternative is the impact or friction forces on the solid materials surface, it has been described that it provides adequate chemical conditions.

Reaction mechanism involving a mechanochemical process is carried out as follows: First, a homolytic rupture of C–C occurs in the PS macromolecule and under atmospheric conditions generates carbon-centered free radical species. Followed by the formation of free radicals which can react with atmospheric oxygen to produce peroxide intermediates. Peroxide species interact with the metal surface of the reactor (which acts as a catalyst) to produce the styrene [70]. Vibratory reactor, stainless steel containers and grinding balls have been employed to generate benzaldehyde and acetophenone through the mechanochemical depolymerization of PS. Under an inert atmosphere (N_2 flow) the production of styrene is mostly favored, however, the incorporation of O_2 flow increases the oxygenated products formation [68].

Another way to reduce the energy necessary for the transformation of PS is through the application of low-cost catalysts (solid), alkaline earth metals salts or oxides have shown high reactivity with the polymer. PS is transformed into benzene, toluene, ethylbenzene, styrene, isopropyl benzene and other (minor) compounds using catalysts such as Mg, MgO and $MgCO_3$ with reaction temperatures <450.0 °C. Under these conditions, it is possible to favor obtaining the styrene monomer yields >50.0% [71].

Another catalytic alternative is the use of metal catalysts supported on mesoporous surfaces such as zeolite, alumina, silica lumina and hydrotalcite [72]. Mixed catalysts described includes calcination of hydrotalcite, mixed metal oxides are formed due to the presence of di- and trivalent metal ions in the starting material, these metal oxide materials possess high surface area and high basicity. Park et al. studied the PS depolymerization employing hydrotalcite with different Mg/Al contents. Reaction was performed at 350.0 °C for 2.0 h; styrene, ethyl benzene and α-methyl styrene were obtained as main reaction products [73]. Vermiculite is other porous material studied for mixed catalyst fabrication. This mineral formed by iron, magnesium and aluminum silicates were used as a catalyst in PS depolymerization reaction at 400.0 °C to production of styrene. The implementation of clay materials as catalysts in the transformation of PS proposes an economical tool, with less energy spending and friendly to the environment [74].

In recent decades, application of supercritical fluids has emerged as an alternative in the chemical transformation of plastics, similar to mechanochemical process, it uses mild reaction conditions that allow it to be a sustainable alternative to thermal treatment. Supercritical water (374.0 °C and 3200.0 psi) is a non-toxic, inexpensive solvent that has allowed the transformation of various polymeric materials.

PS sample and water were place into a supercritical fluid reactor previously purged with argon, the sample is stirred until reaching 380.0 °C. Under these conditions the depolymerization occurs in 15.0 min showing a total conversion. The reaction products forms were styrene monomer, dimers and trimers of styrene, toluene, ethyl benzene and triphenyl benzene [75]. Methanol has been used under supercritical conditions (240.0 °C and 7.9 MPa.) to PS depolymerization, after 15.0 min of reaction a conversion of 92.0% weight was obtained. The exchange of solvents promotes the synthesis of 3-phenyl propanol, 3-phenyl-1-butanol and 1,3-diphenyl propane [76].

On other hand, non-polar solvent allows the dissolution and purification of the depolymerization products. Under this concept, the influence of toluene under supercritical conditions (318.0 °C and 598.0 psi) for PS depolymerization has been evaluated, above 300.0 °C, ethylbenzene, styrene, benzene, among others, are obtained as depolymerization products. 1,3-diphenylpropane is the majority product obtained at temperatures below 300.0 °C, while at temperatures above 350.0 °C this product decomposes into a styrene monomer [77]. Transformation strategy of polymeric waste using supercritical fluids is a promising tool in chemical recycling because of the experimental conditions are environmentally friendly in contrast to thermal treatment.

5 Other Chemical Techniques for PS Depolymerization

Despite achieving depolymerization with the previously described methodologies, they present several disadvantages such as: high reaction temperatures, use of catalysts and solvents [4], therefore, alternative depolymerization techniques have been proposed. Among the alternatives, photocatalytic upcycling presents advantages such as reduced process temperature, reaction rate, and product yields. PS photocatalytic was performed using an iron catalyst under light irradiation to obtain benzoic acid as the majority product, which is a high-value compound that is difficult and expensive to synthesize. A depolymerization yield of 68.0% was obtained under 400.0 nm irradiation, an O_2 atmosphere, and the application iron salts ($FeCl_3$ and $FeBr_3$) as catalysts for the formation of halogen radicals to produce various benzoic compounds (acetophenone and benzoic acid) [78]. Also, photocatalytic recycling in the absence of metal species has been studied, using p-toluene sulfonic acid monohydrate as a catalyst under irradiation at 405.0 nm at air atmosphere to produce formic acid, benzoic acid and acetophenone [18]. The use of microwaves to depolymerize PS presents an effective alternative to conventional heating methods, the application of microwaves increases reaction rates and reduces reaction times and costs [79]. Due to its effectiveness, this methodology has been mostly used in the production of fuels by pyrolysis [80, 81].

Enzymatic depolymerization is other effective method for PS depolymerize because it can gradually break down the polymer chains without the need to implement expensive reactions, unlike conventional methods that require the use of high temperatures and energy [82]. PS is depolymerized using various enzymes such as cytochrome P4500, alkane hydroxylases and monooxygenases. They have been shown to be highly potential in efficiently breaking C–C bonds in PS polymer chains; hydroxylating dioxygenases can break PS side chains to oxidize aromatic rings. Several of these enzymes are produced by microorganisms, and are discarded extracellularly, which in combination with other compounds adhered to PS surface [83]. Decomposition of the polymer begins with the initial cleavage of insoluble macromolecules into smaller fragments as styrene, it can be metabolized through two catalytic pathways: epoxidation of the methyl group or hydroxylation of the benzene ring, yielding products such as benzoic acid, benzaldehyde, acetophenone and ethylbenzene. Some of these products can be used by the bacteria to generate the enzyme [82]. PS was depolymerized using two enzymes, orphan aromatic ring-cleaving dioxygenase and a hydrolase obtained from *Exiguobacterium* sp. RIT 594. It was observed an increase in carboxyl and hydroxyl compounds, as well as non-conjugated double bonds, causing a styrene ring dearomatization. Furthermore, it was identified that the presence of molecular oxygen was a critical parameter for the depolymerization process, obtaining as products 2-hydroxypenta-2,4-dienoate and acrylic acid [84]. However, although the use of enzymes as biocatalysts is a promising area that allows obtaining diverse products and a highly efficient depolymerization under the concept of chemical recycling [85].

6 Conclusion

PS chemical transformation is an innovative process with high recycling efficiency and low energy requirements. Chemical recycling methods emerges as an alternative for the reduction of polystyrene waste and its conversion to different chemical compounds (styrene monomers, benzoic acid, toluene and benzene). On the other hand, it is possible to obtain new materials and morphologies (waste dissolution with organic o natural solvents) and high-value products with diverse chemical applications such as filters or adsorbent materials when PS structural modification by nitration or sulfonation, depolymerization under mechanochemical, photocatalytic, enzymatic or supercritical conditions was applied.

References

1. Sethi B (2016) Methods of recycling. Recycl Polym: Methods, Characterization Appl 55–114. https://doi.org/10.1002/9783527689002.ch3
2. Gutiérrez C, de Haro JC, García MT, Gracia I, de Lucas A, Rodríguez JF (2014) Polystyrene wastes: threat or Opportunity?. In: Jiménez E, Cabañas B, Lefebvre G (eds) Environment, energy and climate change I. The handbook of environmental chemistry, vol 32. Springer, Cham. https://doi.org/10.1007/698_2014_279
3. Grigore ME (2017) Methods of recycling, properties and applications of recycled thermoplastic polymers. Recycling 2(4):24–35. https://doi.org/10.3390/recycling2040024
4. Beghetto V, Sole R, Buranello C, Al-Abkal M, Facchin M (2021) Recent advancements in plastic packaging recycling: a mini-review. Materials 14(17):4782–4806. https://doi.org/10.3390/ma14174782
5. Capricho JC, Prasad K, Hameed N, Nikzad M, Salim N (2022) Upcycling polystyrene. Polymers-Basel 14(22):5010–5058. https://doi.org/10.3390/polym14225010
6. Siddiqui MN, Redhwi HH, Al-Arfaj AA, Achilias DS (2021) Chemical recycling of pet in the presence of the bio-based polymers, pla, phb and pef: a review. Sustainability-Basel 13(19):10528–10554. https://doi.org/10.3390/su131910528
7. Chen H, Wan K, Zhang Y, Wang Y (2021) Waste to wealth: chemical recycling and chemical upcycling of waste plastics for a great future. Chemsuschem 14(19):4123–4136. https://doi.org/10.1002/cssc.202100652
8. Liu Y, Qian J, Wang J (2000) Pyrolysis of polystyrene waste in a fluidized-bed reactor to obtain styrene monomer and gasoline fraction. Fuel Process Technol 63(1):45–55. https://doi.org/10.1016/S0378-3820(99)00066-1
9. Wilk V, Hofbauer H (2013) Conversion of mixed plastic wastes in a dual fluidized bed steam gasifier. Fuel 107:787–799. https://doi.org/10.1016/j.fuel.2013.01.068
10. Eloka-Eboka AC, Maroa S (2021). Advanced and sustainable biodiesel fuels: technologies and applications. Sustainable Biofuels 131–161. https://doi.org/10.1016/B978-0-12-820297-5.00005-0

11. Kijo-Kleczkowska A, Gnatowski A (2022) Recycling of plastic waste, with particular emphasis on thermal methods. Energies 15(6):2114–2135. https://doi.org/10.3390/en15062114
12. Maafa IM (2021) Pyrolysis of polystyrene waste: a review. Polymers-Basel 13(2):225–255. https://doi.org/10.3390/polym13020225
13. European Parliament (2008) Directive 2008/98/ec of the European parliament and of the council of 19 November 2008 on waste and repealing certain directives. Official J European Union, 312. https://eur-lex.europa.eu/legal-content/EN/TXT/PDF/?uri=CELEX:32008L0098&from=EN
14. Tanaka M, Taguchi Y, Sawatari N, Saito N (2006) Preparation of composite particles coated with two kinds of solid powders by semi-chemical recycle method. J Mater Sci 41:4215–4220. https://doi.org/10.1007/s10853-006-6273-9
15. Achilias DS, Giannoulis A, Papageorgiou GZ (2009) Recycling of polymers from plastic packaging materials using the dissolution-reprecipitation technique. Polym Bull 63:449–465. https://doi.org/10.1007/s00289-009-0104-5
16. Escamilla-Lara KA, Lopez-Tellez J, Rodriguez JA (2023) Adsorbents obtained from recycled polymeric materials for retention of different pollutants: a review. Chemosphere 335:139159. https://doi.org/10.1016/j.chemosphere.2023.139159
17. Gautam B, Huang MR, Lin CC, Chang CC, Chen JT (2023) A viable approach for polymer upcycling of polystyrene (styrofoam) wastes to produce high value predetermined organic compounds. Polym Degrad Stabil 217:110528. https://doi.org/10.1016/j.polymdegradstab.2023.110528
18. Huang Z, Shanmugam M, Liu Z, Brookfield A, Bennett EL, Guan R, Xiao J (2022) Chemical recycling of polystyrene to valuable chemicals via selective acid-catalyzed aerobic oxidation under visible light. J Am Chem Soc 144(14):6532–6542. https://doi.org/10.1021/jacs.2c01410
19. Mumbach GD, Bolzan A, Machado RAF (2020) A closed-loop process design for recycling expanded polystyrene waste by dissolution and polymerization. Polymer 209:122940. https://doi.org/10.1016/j.polymer.2020.122940
20. Damayanti D, Saputri DR, Marpaung DSS, Yusupandi F, Sanjaya A, Simbolon YM, Wu HS (2022) Current prospects for plastic waste treatment. Polymers-Basel 14(15):3133–3162. https://doi.org/10.3390/polym14153133
21. Lopez-Tellez J, Rodriguez JA, Miranda JM, Mondragon AC, Ibarra IS (2022) Determination of polycyclic aromatic hydrocarbons in aqueous samples using recycled polystyrene for pipette tip-solid phase extraction followed by HPLC-FLD. Int J Environ An Ch 1–10. https://doi.org/10.1080/03067319.2022.2086051
22. Fadil F, Affandi NDN, Misnon MI, Bonnia NN, Harun AM, Alam MK (2021) Review on electrospun nanofiber-applied products. Polymers-Basel 13(13):2087–2116. https://doi.org/10.3390/polym13132087
23. Fan L, An J, Gao J, Cui Y, Dong Z (2022) Determination of atypical antipsychotics in human plasma by UPLC-UV with polystyrene nanofibers as a solid-phase extraction sorbent. RSC Adv 12(25):16194–16202. https://doi.org/10.1039/D2RA02457H
24. Ríos-Gómez J, Lucena R, Cárdenas S (2017) Paper supported polystyrene membranes for thin film microextraction. Microchem J 133:90–95. https://doi.org/10.1016/j.microc.2017.03.026
25. Mehmandost N, Soriano ML, Lucena R, Goudarzi N, Chamjangali MA, Cardenas S (2019) Recycled polystyrene-cotton composites, giving a second life to plastic residues for environmental remediation. J Environ Chem Eng 7(5):103424. https://doi.org/10.1016/j.jece.2019.103424
26. Munir MM, Adrian M, Burhanuddin M, Iskandar F (2022) Fabrication and structure optimization of expanded polystyrene (EPS) waste fiber for high-performance air filtration. Powder Technol 402:117357. https://doi.org/10.1016/j.powtec.2022.117357
27. Martins TR, Costa PS, Bertuol DA, Aguiar ML, Tanabe EH (2022) Development of recycled expanded polystyrene nanofibers modified by chitosan for the removal of lead (II) from water. Metals-Basel 12(8):1334–1351. https://doi.org/10.3390/met12081334
28. Santos MMDM, Duarte MMMB, Nascimento GED, Souza NBGD, Rocha ORSD (2019) Use of TiO_2 photocatalyst supported on residues of polystyrene packaging and its applicability on the removal of food dyes. Environ Technol 40(12):1494–1507. https://doi.org/10.1080/09593330.2017.1423396

29. Shin C (2006) Filtration application from recycled expanded polystyrene. J Colloid Interf Sci 302(1):267–271. https://doi.org/10.1016/j.jcis.2006.05.058
30. Liu F, Song D, Huang X, Xu H (2016) Electrospun polystyrene nanofibers as a novel adsorbent to transfer an organic phase from an aqueous phase. J Sep Sci 39(7):1326–1330. https://doi.org/10.1002/jssc.201501182
31. Ghambari H, Reyes-Gallardo EM, Lucena R, Saraji M, Cárdenas S (2017) Recycling polymer residues to synthesize magnetic nanocomposites for dispersive micro-solid phase extraction. Talanta 170:451–456. https://doi.org/10.1016/j.talanta.2017.04.026
32. An J, Dong Z, Zhang W, Yan Y, Kang W, Lian K (2021) Development of a simple nanofiber-based solid phase extraction procedure coupled with high performance liquid chromatography analysis for the quantification of eight sedative-hypnotic drugs in human urine samples. Microchem J 168:106475. https://doi.org/10.1016/j.microc.2021.106475
33. Isık T, Demir MM (2018) Tailored electrospun fibers from waste polystyrene for high oil adsorption. SM&T 18:e00084. https://doi.org/10.1016/j.susmat.2018.e00084
34. Zali S, Jalali F, Es-Haghi A, Shamsipur M (2015) Electrospun nanostructured polystyrene as a new coating material for solid-phase microextraction: application to separation of multipesticides from honey samples. J Chromatogr B 1002:387–393. https://doi.org/10.1016/j.jchromb.2015.07.061
35. Hashemifard N, Shariati S (2016) Electrospun polystyrene nanofiber as an adsorbent for solid-phase extraction of disulfine blue from aqueous samples. Arab J Sci Eng 41:2487–2492. https://doi.org/10.1007/s13369-015-1968-3
36. Saleh HM, Khalil AM (2022) Plastic films based on waste expanded polystyrene loaded with bagasse powder for packaging applications. Egypt J Chem 65(11):323–329. https://doi.org/10.21608/ejchem.2022.123154.5507
37. Gutierrez-Velasquez EI, Monteiro SN, Colorado HA (2022) Characterization of expanded polystyrene waste as binder and coating material. Case Stud Constr Mater 16:e00804. https://doi.org/10.1016/j.cscm.2021.e00804
38. Lopez-Tellez J, Ibarra IS, Cruz-Borbolla J, Vega M, Rodriguez JA (2023) Retention and determination of polycyclic aromatic hydrocarbons from urban air based on recycled polyurethane foam modified with expanded polystyrene. Polycycl Aromat Comp 43(10):9347–9359. https://doi.org/10.1080/10406638.2022.2162931
39. Zulfi A, Munir MM, Hapidin DA, Rajak A, Edikresnha D, Iskandar F, Khairurrijal K (2018) Air filtration media from electrospun waste high-impact polystyrene fiber membrane. Mater Res Express 5(3):035049. https://doi.org/10.1088/2053-1591/aab6ef
40. Rajak A, Hapidin DA, Iskandar F, Munir MM, Khairurrijal K (2019) Controlled morphology of electrospun nanofibers from waste expanded polystyrene for aerosol filtration. Nanotechnology 30(42):425602. https://doi.org/10.1088/1361-6528/ab2e3b
41. Winterton N (2021) The green solvent: a critical perspective. Clean Techn Environ Policy 23:2499–2522. https://doi.org/10.1007/s10098-021-02188-8
42. Gil-Jasso ND, Segura-González MA, Soriano-Giles G, Neri-Hipolito J, López N, Mas-Hernández E, Ballesteros-Rivas MF (2019) Dissolution and recovery of waste expanded polystyrene using alternative essential oils. Fuel 239:611–616. https://doi.org/10.1016/j.fuel.2018.11.055
43. Gil-Jasso ND, Giles-Mazón EA, Soriano-Giles G, Reinheimer EW, Varela-Guerrero V, Ballesteros-Rivas MF (2022) A methodology for recycling waste expanded polystyrene using flower essential oils. Fuel 307:121835. https://doi.org/10.1016/j.fuel.2021.121835
44. Hattori K, Naito S, Yamauchi K, Nakatani H, Yoshida T, Saito S, Miyakoshi T (2008) Solubilization of polystyrene into monoterpenes. Adv Polym Tech 27(1):35–39. https://doi.org/10.1002/adv.20115
45. Hattori K (2015). Recycling of expanded polystyrene using natural solvents. Recycling materials based on environmentally friendly techniques, 1–15. InTech. https://doi.org/10.5772/59156
46. Ozdemir NC, Yel E (2023) Synthesis of a new flocculant from waste polystyrene: plastic recycling industry wastewater treatability. Water Air Soil Pollut 234:88. https://doi.org/10.1007/s11270-023-06104-2

47. Santos ASF, Teixeira BAN, Agnelli JAM, Manrich S (2005) Characterization of effluents through a typical plastic recycling process: an evaluation of cleaning performance and environmental pollution. Resour Conserv Recycl 45(2):159–171. https://doi.org/10.1016/j.resconrec.2005.01.011
48. Bekri-Abbes I, Bayoudh S, Baklouti M (2006) Converting waste polystyrene into adsorbent: potential use in the removal of lead and cadmium ions from aqueous solution. J Polym Environ 14:249–256. https://doi.org/10.1007/s10924-006-0018-3
49. Sułkowski WW, Nowak K, Sułkowska A, Wolińska A, Bajdur WM, Pentak D, Mikuła B (2010) Chemical recycling of polystyrene. Sulfonation with different sulfonation agents. Mol Cryst Liq Cryst 523(1):218–790. https://doi.org/10.1080/15421401003720140
50. De Leon-Condes CA, Roa-Morales G, Martinez-Barrera G, Balderas-Hernandez P, Menchaca-Campos C, Urena-Nunez F (2019) A novel sulfonated waste polystyrene/iron oxide nanoparticles composite: green synthesis, characterization and applications. J Environ Chem Eng 7(1):102841. https://doi.org/10.1016/j.jece.2018.102841
51. Dardeer HM, Toghan A (2021) A novel route for the synthesis of pseudopolyrotaxane containing γ-Cyclodextrin based on environmental waste recycling. J Mol Struct 1227:129707. https://doi.org/10.1016/j.molstruc.2020.129707
52. Mahmoud ME, Abdou AE, Ahmed SB (2016) Conversion of waste styrofoam into engineered adsorbents for efficient removal of cadmium, lead and mercury from water. ACS Sustain Chem Eng 4(3):819–827. https://doi.org/10.1021/acssuschemeng.5b01149
53. Sułkowski WW, Nowak K, Sułkowska A, Mikuła B, Wierzba P (2013) The conditions of cationic exchange with the use of recycling polystyrene derivative, the product of sulfonation by silica sulfuric acid. J Appl Polym Sci 128(5):2611–2617. https://doi.org/10.1002/app.38429
54. Ruziwa D, Chaukura N, Gwenzi W, Pumure I (2015) Removal of Zn^{2+} and Pb^{2+} ions from aqueous solution using sulphonated waste polystyrene. J Environ Chem Eng 3(4):2528–2537. https://doi.org/10.1016/j.jece.2015.08.006
55. Chang SH, Lu CC, Lin CW, Wang KS, Lee MW, Liu SH (2022) Waste expanded polystyrene modified with H_2SO_4/biodegradable chelating agent for reuse: as a highly efficient adsorbent to remove fluoroquinolone antibiotic from water. Chemosphere 288:132619. https://doi.org/10.1016/j.chemosphere.2021.132619
56. Tang X, Din C, Yu S, Liu Y, Luo H, Zhang D, Chen S (2022) Synthesis of dielectric polystyrene via one-step nitration reaction for large-scale energy storage. Chem Eng J 446:137281. https://doi.org/10.1016/j.cej.2022.137281
57. Merchán-Arenas DR, Murcia-Patiño AF (2021) Synthesis of polyamino styrene from post-consumption expanded polystyrene and analysis of its CO_2 scavenger capacity. Int J Environ Sci Technol 18(9):2519–2532. https://doi.org/10.1007/s13762-020-03009-z
58. Pu Y, Xie Z, Gong X, Yan Y, Zhang J (2021) Study on removal of organic dyes by Fe_3O_4/amidation modified waste polystyrene composites. Environ Technol Innov 23:101732. https://doi.org/10.1016/j.eti.2021.101732
59. Siyal AN, Memon SQ, Parveen S, Soomro A, Khaskheli MI, Khuhawar MY (2013) Chemical recycling of expanded polystyrene waste: synthesis of novel functional polystyrene-hydrazone surface for phenol removal. J Chem-Ny 2013(1):842435. https://doi.org/10.1155/2013/842435
60. Maleki F, Ghaemi A, Mir Mohamad Sadeghi G (2023) Synthesis and characterization of waste Styrofoam hypercrosslinked polymer as an adsorbent for CO_2 capture. Environ Prog Sustain 42(1):e13954. https://doi.org/10.1002/ep.13954
61. Pu Y, Xie Z, Ye H, Shi W (2021) Amidation modified waste polystyrene foam as an efficient recyclable adsorbent for organic dyes removal. Water Sci Technol 83(9):2192–2206. https://doi.org/10.2166/wst.2021.129
62. Nair CR, Manshad PK, Ashir AM, Athul S (2020) Synthesis of 3-carbonoyl acrylic acid-functionalized polystyrene and an insight in to its role in cross linking and grafting of polystyrene on to natural rubber. Eur Polym Jl 131:109688. https://doi.org/10.1016/j.eurpolymj.2020.109688
63. Li W, Xie Z, Xue S, Ye H, Liu M, Shi W, Liu Y (2021) Studies on the adsorption of dyes, Methylene blue, Safranin T, and Malachite green onto Polystyrene foam. Sep Purif Technol 276:119435. https://doi.org/10.1016/j.seppur.2021.119435

64. Wang Z, Bai R, Ting YP (2008) Conversion of waste polystyrene into porous and functionalized adsorbent and its application in humic acid removal. Ind Eng Chem Res 47(6):1861–1867. https://doi.org/10.1021/ie070583m
65. Fu Z, Mohamed IM, Li J, Liu C (2019) Novel adsorbents derived from recycled waste polystyrene via cross-linking reaction for enhanced adsorption capacity and separation selectivity of CO_2. J Taiwan Inst Chem E 97:381–388. https://doi.org/10.1016/j.jtice.2019.01.014
66. Dong X, Akram A, Comesaña-Gándara B, Dong X, Ge Q, Wang K, Lau CH (2020) Recycling plastic waste for environmental remediation in water purification and CO_2 capture. ACS Appl Polym Mater 2(7):2586–2593. https://doi.org/10.1021/acsapm.0c00224
67. Chen D, Xie Z, Ye H, Li W, Shi W, Liu Y (2022) Upcycling of expanded polystyrene waste: amination as adsorbent to recover Eriochrome Black T and Congo red. Sep Purif Technol 289:120669. https://doi.org/10.1016/j.seppur.2022.120669
68. Chang Y, Blanton SJ, Andraos R, Nguyen VS, Liotta CL, Schork FJ, Sievers C (2023) Kinetic phenomena in mechanochemical depolymerization of poly (styrene). ACS Sustain Chem Eng 12(1):178–191. https://doi.org/10.1021/acssuschemeng.3c05296
69. Yao L, King J, Wu D, Ma J, Li J, Xie R, Peng Z (2022) Non-thermal plasma-assisted rapid hydrogenolysis of polystyrene to high yield ethylene. Nat Commun 13(1):885. https://doi.org/10.1038/s41467-022-28563-7
70. Balema VP, Hlova IZ, Carnahan SL, Seyedi M, Dolotko O, Rossini AJ, Luzinov I (2021) Depolymerization of polystyrene under ambient conditions. New J Chem 45(6):2935–2938. https://doi.org/10.1039/D0NJ05984F
71. Shah J, Jan MR, Adnan (2014) Conversion of waste polystyrene through catalytic degradation into valuable products. Korean J Chem Eng 31:1389–1398. https://doi.org/10.1007/s11814-014-0016-4
72. Marczewski M, Kamińska E, Marczewska H, Godek M, Rokicki G, Sokołowski J (2013) Catalytic decomposition of polystyrene. the role of acid and basic active centers. Appl Catal B: Environ 129:236–246. https://doi.org/10.1016/j.apcatb.2012.09.027
73. Park SJ, Kang SH, Min HK, Choi YH, Lee JW (2021) Depolymerization of polystyrene over Mgx-Aly-O catalysts derived from hydrotalcites: effect of Mg/Al ratio on the basicity and catalytic performance. Mol Catalysis 506:111546. https://doi.org/10.1016/j.mcat.2021.111546
74. Valášková M, Madejová J, Inayat A, Matějová L, Ritz M, Martaus A, Leštinský P (2020) Vermiculites from Brazil and Palabora: structural changes upon heat treatment and influence on the depolymerization of polystyrene. Appl Clay Sci 192:105639. https://doi.org/10.1016/j.clay.2020.105639
75. Kwak H, Shin HY, Bae SY, Kumazawa H (2006) Characteristics and kinetics of degradation of polystyrene in supercritical water. J Appl Polym Sci 101(1):695–700. https://doi.org/10.1002/app.23896
76. Shin HY, Bae SY (2008) Thermal decomposition of polystyrene in supercritical methanol. J Appl Polym Sci 108(6):3467–3472. https://doi.org/10.1002/app.27960
77. Saha N, Banivaheb S, Reza MT (2022) Towards solvothermal upcycling of mixed plastic wastes: depolymerization pathways of waste plastics in sub-and supercritical toluene. Energy Convers Manag X 13:100158. https://doi.org/10.1016/j.ecmx.2021.100158
78. Parkatzidis K, Wang HS, Anastasaki A (2024) Photocatalytic upcycling and depolymerization of vinyl polymers. Angew Chem 136(19):e202402436. https://doi.org/10.1002/ange.202402436
79. Gautam B, Tsai TH, Chen JT (2024) Towards sustainable solutions: a review of polystyrene upcycling and degradation techniques. Polym Degrad Stabil 110779. https://doi.org/10.1016/j.polymdegradstab.2024.110779
80. Fan S, Zhang Y, Cui L, Xiong Q, Maqsood T (2023) Conversion of polystyrene plastic into aviation fuel through microwave-assisted pyrolysis as affected by iron-based microwave absorbents. ACS Sustain Chem Eng 11(3):1054–1066. https://doi.org/10.1021/acssuschemeng.2c05880
81. Yang C, Shang H, Li J, Fan X, Sun J, Duan A (2023) A review on the microwave-assisted pyrolysis of waste plastics. Processes 11(5):1487–1509. https://doi.org/10.3390/pr11051487

82. Hou L, Majumder ELW (2021) Potential for and distribution of enzymatic biodegradation of polystyrene by environmental microorganisms. Materials 14(3):503–523. https://doi.org/10.3390/ma14030503
83. Kim HR, Lee HM, Yu HC, Jeon E, Lee S, Li J, Kim DH (2020) Biodegradation of polystyrene by Pseudomonas sp. isolated from the gut of superworms (larvae of Zophobas atratus). Environ Sci Technol 54(11):6987–6996. https://doi.org/10.1021/acs.est.0c01495
84. Parthasarathy A, Miranda RR, Eddingsaas NC, Chu J, Freezman IM, Tyler AC, Hudson AO (2022) Polystyrene degradation by Exiguobacterium sp. RIT 594: preliminary evidence for a pathway containing an atypical oxygenase. Microorganisms 10(8):1619–1637. https://doi.org/10.3390/microorganisms10081619
85. Kim YB, Kim S, Park C, Yeom SJ (2024). Biodegradation of polystyrene and systems biology-based approaches to the development of new biocatalysts for plastic degradation. Curr Opin Syst Biol, 100505. https://doi.org/10.1016/j.coisb.2024.100505

Components of Expanded Polystyrene Foams

Edwin A. Giles-Mazón, Victor Varela-Guerrero, and Maria F. Ballesteros-Rivas

Abstract Expanded polystyrene (EPS) is an irreplaceable material due to unique properties like lightweight structure, impact tolerant properties and thermal insulation making it essential for daily lives applications. It can be found in applications that take full advantage of these properties such as packing, construction and many others. This chapter explores the chemical composition of EPS, starting from the chemical components of the styrene monomer and its transformation into polystyrene (PS) and EPS. In the process of obtaining EPS, some additives are involved, which are substances that help to obtain EPS with certain characteristics and can also help to improve certain properties that are needed to efficiently use EPS in certain applications. In addition, the path to reach EPS of different densities is explored. EPS foams can be classified based on their density following international norms and standards. The norms ensure the quality of EPS products and define the physical parameters based on their density values. The value of density helps to know what properties each type of EPS foam has, for example, rigidity, mechanical resistance, thermal insulation and resistance to humidity are enhanced when the density values are high. Finally, this chapter highlights the close relationship that exists between the physical properties resulting from each type of EPS, and the apolar nature of styrene and its distribution between the air cells of EPS.

Keywords Styrene · Polystyrene · Blowing agents · Expandable polystyrene · Expanded polystyrene

---

E. A. Giles-Mazón · V. Varela-Guerrero (✉) · M. F. Ballesteros-Rivas (✉)
Facultad de Química, Universidad Autónoma del Estado de México, Toluca de Lerdo, México
e-mail: vvarelag@uaemex.mx

M. F. Ballesteros-Rivas
e-mail: mfballesterosr@uaemex.mx

V. Varela-Guerrero · M. F. Ballesteros-Rivas
Centro Conjunto de Investigación en Química Sustentable, UAEM-UNAM, Toluca de Lerdo, México

A. Khan et al. (eds.), *Uses and Products of Recycled Expanded Polystyrene Foam Wastes*, Sustainable Materials and Technology,
https://doi.org/10.1007/978-981-96-5618-9_5

1 Introduction

Expanded polystyrene (EPS) foam is one of the most common plastics used in our daily lives [1]. If the use of EPS foam increased, its waste has also increased proportionality and the high consumption EPS represents a global environmental problem [2]. EPS is used in a plethora of applications like packaging, construction and household appliances due to their low weight, rigidity, high impact resistance, heat insulation capabilities and moisture resistance property, making it difficult to replace [3]. The physical characteristics of EPS foams are due to their chemical components and their three-dimensional arrangement obtain by the additives added on EPS production process [4, 5]. EPS is made of 98% air and 2% polystyrene (PS) particles [2, 6]. PS is a thermoplastics substance (polymer that can be melted and molted) that is exceptional for reuse, redesign and recycling in the circular economy [1, 7]. In this chapter, we will explore the fundamental components of EPS foams, to understand better their chemical and physical properties that make EPS foams one of the most consumed plastics. In addition, understanding the chemical of EPS components leads to new methods of producing and recycling polystyrene considered sustainability [8]. Also, knowing the chemical behavior of the EPS components on determinate recycling process is essential to predict the yield of a given value-added products results of the process.

2 Components of EPS Foam and Chemical Properties

EPS is a polymeric structure based on styrene monomer [4]. Styrene is found in nature in plants, bacteria and fungi, where it participates in some biochemical processes. However, to meet the market demand, styrene for plastic production is obtained by organic synthesis and is a petroleum-derived [3]. Therefore, analyzing this molecule considering each of its chemical components allows us to understand in greater detail the close relationship that exists between the arrangement and chemical properties of the styrene chains and the EPS physical properties.

2.1 Styrene (C_8H_8)

Styrene has a molecular formula of C_8H_8. It is a hydrocarbon made up of eight hydrogen atoms distributed and linked between eight carbon atoms [9]. This composition of only hydrogen and carbon atoms gives it the property of being a non-polar and hydrophobic molecule, which allows it to be impermeable to water and resistant to humidity, properties that are also maintained in its polymeric form, both in PS and EPS [4]. The molecular formula refers to the number and type of atoms present in a molecule, therefore, to obtain more information about the structure and chemical behavior of the molecule in question, the developed chemical formula details the mode in which the atoms are linked to each other within the molecule. Figure 1a shows the developed chemical formula from styrene [10]. Two functional groups are distinguished in the molecule: the phenyl group (C_8H_5-) which corresponds to the benzene ring with one less hydrogen and, as a substituent, the vinyl group ($-CH = CH_2$), the corresponding radical of ethylene molecule ($H_2C = CH_2$) [11]. In fact, the names *Phenylethylene* and *Vinylbenzene* are synonyms of styrene, and they denote the presence of these functional groups. Functional groups are a certain set of atoms that have similarity in chemical properties and reactivity, so that they define the interaction of a molecule with others. In the case of styrene, its reactive properties are defined by the vinyl group, and it is the one that directly participates in the polymerization of the monomers to form PS (Fig. 1b and c). The phenyl group consists of a highly non-polar aromatic ring due to the delocalized distribution of its electrons, which gives rise to an electron cloud that extends throughout the carbon ring structure [12]. This configuration has a direct influence on the stability of the styrene molecule and the non-polar properties of the molecule, making it insoluble in polar solvents such as water and soluble in non-polar solvents. A phenomenon that is exploited in chemical recycling processes by dissolving PS and EPS or in adhesive manufacturing processes [2, 13, 14].

2.2 *Polystyrene (PS)*

PS has the next molecular formula: $(C_8H_8)_n$, where n indicates the number of times the styrene monomer is repeated. Figure 1b shows the structural formula of PS. It is observed that for polymerization to occur, the double bonds of the vinyl group of styrene are transformed into single bonds that allow the continuous addition of styrene monomers through the formation of new carbon–carbon bonds, releasing energy in the form of heat (exothermic process) and achieving monomeric elongation in the form of linear chains (Fig. 1c) [12].

Polystyrene is divided into two types: solid PS and expanded polystyrene (EPS). Commercial polystyrene is an amorphous material with a molecular weight $Mw = 100.000 \pm 400.000$ Daltons [4]. Molecular weight is the mass of a molecule and is defined as the sum of the atomic masses of the atoms that make up the molecule. For polymers, such as PS, their weight is defined by the average of the masses of the long chains of monomeric units [9].

The techniques for obtaining commercial polystyrene are different for solid PS and EPS. In general, if PS is to be foamed (EPS), they are based on a suspension process because the blowing agent can be introduced during polymerization, or on a bulk polymerization process for solid PS [4, 5]. After the polymer melt is pelletized, in small beads, the diluents are recovered and recycled [9].

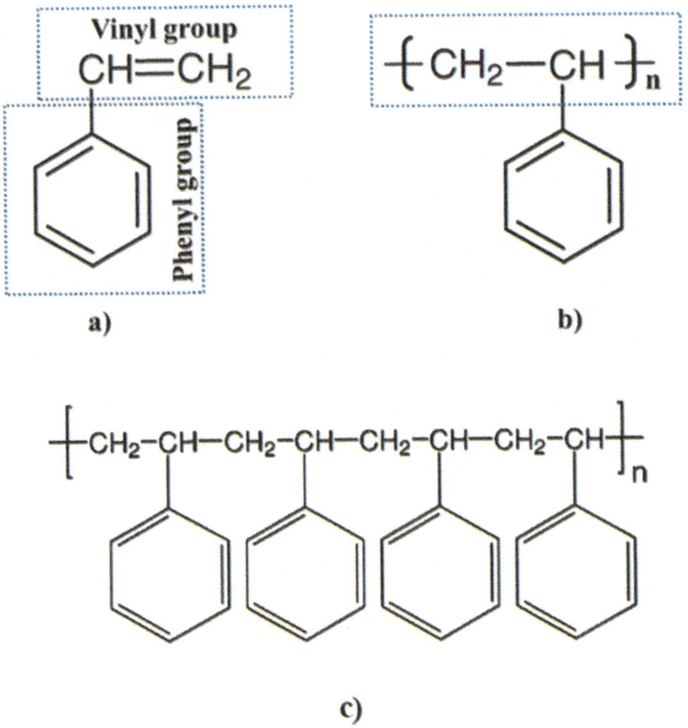

Fig. 1 Molecular structure of styrene (**a**) and polystyrene (**b**) (**c**)

3 Conversation of Expandable PS Beads and EPS Foams

In the process of obtaining EPS the small beads formed by the polymerization of styrene are expandable PS beads. The production of high quantities of expandable beads of polystyrene involves the polymerization where the granules are formed and the addition of pentane and/or other blowing agents (like butane, or carbon dioxide), which diffuse into the granule [4–6]. Conversion of expandable polystyrene to expanded polystyrene is carried out in three stages: (1) Pre-expansion, (2) intermediate mature and stabilization, and (3) formulation and molding (Fig. 2) [1, 3, 6].

3.1 Pre-Expansion

The expandable PS beads are put on machines called pre-expanders, where the temperatures increase between 80 and 100 °C. At higher temperatures, it may lose its physical properties due to temperature degradation. These temperatures rise the

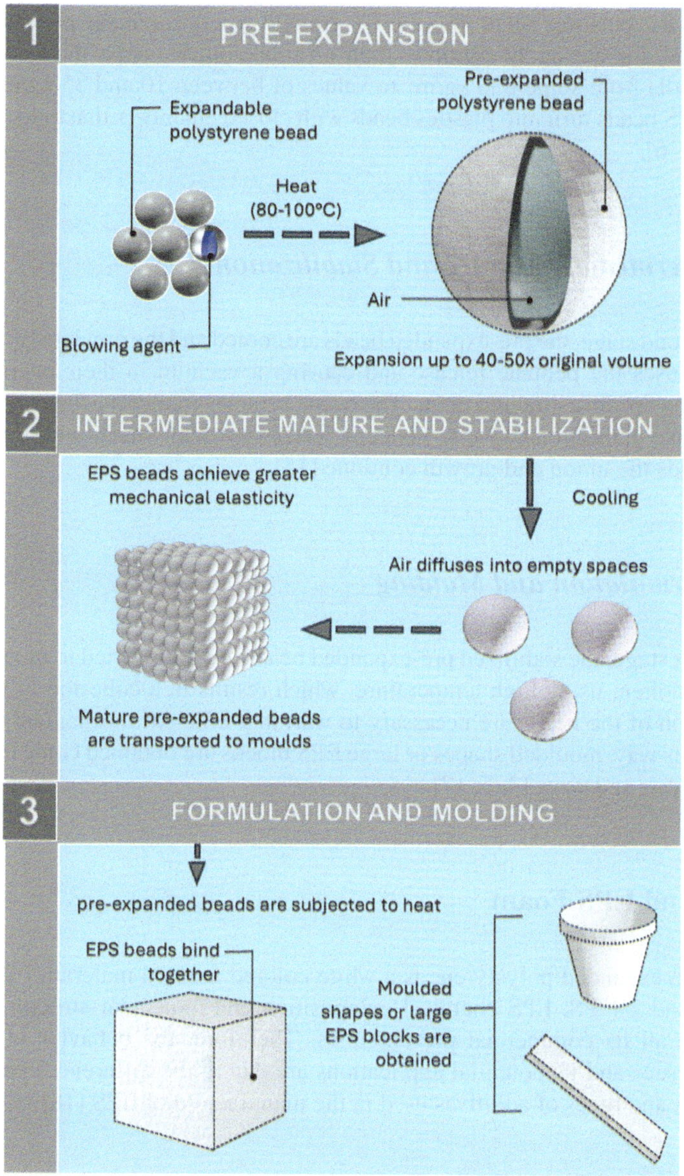

Fig. 2 The conversion process of expandable polystyrene to expanded polystyrene

kinetic of the blowing agent into the PS granule where the inner pressure softens the material, increasing 40–50 times their original volume. Also, the density of the material falls from some 630 kg/m^3 to values of between 10 and 35 kg/m^3 due the compact PS beads turn into plastics beads with closed chambers that hold air in their interior [3–6].

3.2 *Intermediate Mature and Stabilization*

On the second stage, the pre-expanded beads are cooled and the previously exposition to heat causes the pentane release and causing a vacuum in their interior that is balanced by air diffusing into them. The beads achieve greater mechanical elasticity and improve expansion capacity hence be processed with steam and/or hot air on block molds the union and growth continue [15].

3.3 *Formulation and Molding*

On the last stage, the stabilized pre-expanded beads are transported to moulds where they are molten, using high temperature, which results in a cohesion of the beads. Good union of the beads are necessary to warranty efficient mechanical properties [16]. In this way, moulded shapes or large EPS blocks are obtained come in a variety of geometries and sizes [3, 5, 17].

4 PS and EPS Foam

In general, expanded polystyrene is a white-colored foamed material. EPS foam is 98% air and 2% PS. EPS chemical composition and molecular structure are very similar in all its commercial presentations. Therefore, the behavior of its physical properties and its potential applications are due to the differences between the treatments and types of additives used in the manufacture of EPS [18].

4.1 *Physical Properties of EPS Foams and Density*

The physical properties are primarily determined by the density of EPS. EPS density is measured in kilograms per cubic meter (kg/m^3). The higher the value, the greater the number of PS chains per cubic meter and the lower the percentage of air cells. In addition, this translates into greater resistance to compression (mechanical resistance), greater rigidity and even greater resistance to moisture because the water

Table 1 EPS classification according to its density [23]

Class	Average density (Kg/m^3)	Compressive strength (kPa)	Moisture resistance ($\mu g/m^2 s$)
L	11	50	710
SL	13.5	70	630
S	16	85	580
M	19	105	520
H	24	135	460
VH	28	165	400

retention capacity is lower due to the more compact arrangement of the PS chains (Table 1) [19]. Therefore, through density, information can be obtained about the physical properties and probable applications of the different EPS foam. Therefore, the classifications of the types of EPS foam are based on the density values and supported by international standards and norms that define the physical and mechanical parameters of EPS [20, 21]. In addition, they ensure the quality of EPS in each country or region of the world [22].

4.2 *Blowing Agents*

In general, the density and therefore the type of PS foam depends on the expansion conditions and the type of blowing agent used during its production process [18]. As mentioned above, the expandable PS beads are polymerized while they are impregnated with a blowing agent like pentane, butane, or carbon dioxide [4, 15, 16]. The blowing agent added is responsible for generating the cells within the material by expanding the PS granules. These cells fill with air by diffusion giving it characteristic light structure.

4.3 *Pentane (C_5H_{12})*

Pentane is a hydrocarbon solvent coming directly from natural gas and crude oil [3]. Pentane is a blowing agent used to expand polystyrene beads during the EPS production process. Pentane is a slightly volatile liquid at room temperature, it not only foams the material, but also remains to some extent in the foam cells [4]. EPS that has been expanded with pentane can be more flammable, necessitating the use of flame-retardant additives in some cases. Pentanes are cheap, efficient, and provide good expansion due to their size, so larger cells can be generated in the EPS [24, 25].

4.4 Carbon Dioxide (CO_2)

Carbon dioxide is a colorless gas composed of one carbon atom and two oxygen atoms (CO_2). It can be used as a blowing agent for the expansion of PS beads; however, it is used to a lesser extent than pentane [26]. The main advantages of CO_2 over pentane are that it is a non-flammable gas, and that it can be obtained from industrial processes, avoiding generating additional emissions [27]. EPS expanded with CO_2 tends to have smaller cells and therefore a denser structure than those obtained using pentane. In addition, because the size of the CO_2 molecule is smaller than the pentane molecule, the expansion is less efficient [28].

4.5 Additives

In the production of EPS, various additives are added to improve the properties of the material or adjust its final characteristics, such as fire resistance, UV stability, or mechanical durability (blowing agents have an important role on these properties) [29]. As mentioned above, EPS may still have some residual pentane, the pentane is a liquid inflammable, and it is common to find EPS with flame retardant substances such as tetrabromobisphenol A (TBBPA) o hexabromocyclododecane (HBCDD) [1, 30, 31]. Bromine atoms on the chemical structure of these flame-retardant substances inhibit combustion. However, due to the environmental damage caused by HBCDD, it is being replaced by more sustainable flame retardants, since this compound is still present in the recycling process. To improve stability against UV radiation, substituted benzophenones, phenols, inorganic and organometallic additives are often added to protect EPS from degradation by exposure to UV radiation and prevent EPS deterioration over time and oxidation [32–34] Inorganic additives such as iron oxide (Fe_2O_3) are also often added, as pigments. Some organic dyes interact directly with the styrene units from the EPS, so it is easily stained. Titanium dioxide (TiO_2) is also often added, like a photocatalyzer to transform organic pollutants into goods [35, 36]. Or even to improve its antimicrobial properties, silver salts are added [37]. Just as a certain density determines its application, the additives added in the EPS production process will depend on the desired application.

5 Chemical Properties of EPS

Considering again the example of moisture resistance, the denser the EPS foam, the more compacted its PS chains are to prevent water or water vapor from entering. From a chemical point of view, the non-polar nature of PS, given by the styrene monomers, makes EPS insoluble in water and other polar solvents such as alcohols, but vulnerable to non-polar solvents [2, 13, 38, 39]. The property of resisting microbial growth by

PS and EPS is partly related to moisture resistance [3, 6]. Since if moisture does not accumulate, the conditions for the growth of microorganisms such as bacteria and fungi are not given, and this makes EPS products safe and can be used as packaging products intended for the food or pharmaceutical area. In addition, if it is required to include an inhibitory effect on bacterial growth, EPS can be coated with silver salts or other antibacterial organic materials [37]. Also due to the non-polar nature of PS, PS has low gas and liquid permeability. In addition, EPS is chemically stable to weak acids and bases, meaning it maintains its intact structure after prolonged exposure to these substances [3, 6]. For this reason, it does not react when exposed to everyday products or substances, making EPS products suitable for packaging and preserving food. For example, when used to store fruit, specifically some citrus fruits, they help retain vitamin C, preventing its loss while on the shelf [40].

6 Conclusion

Expanded polystyrene (EPS) is a versatile material that can be used in many applications, making it one of the most widely used plastics in the world. Obtaining EPS involves chemical processes where various substances interact with the monomeric chemical components of polystyrene. Among these substances are additives such as blowing agents that play a crucial role in generating cellular structures and directly affect the final density, and therefore also its mechanical and thermal properties. Other additives give EPS specific characteristics such as fire resistance, greater durability when exposed to ultraviolet light and protection against oxidation. Thus, the range of application areas of EPS can be expanded, allowing its use in industries, construction, packaging, thermal insulation, sound insulation, among many other applications.

References

1. Sharma C, Qanungo K (2023) An overview: recycling of expanded polystyrene foam. AIP Conf Proc. 2535(1):030019. https://doi.org/10.1063/5.0111982
2. Gil-Jasso ND, Segura-González MA, Soriano-Giles G, Neri-Hipolito J, López N, Mas-Hernández E, et al (2019) Dissolution and recovery of waste expanded polystyrene using alternative essential oils. Fuel 239:611–616. Available from: https://www.sciencedirect.com/science/article/pii/S0016236118319410
3. BPF EPS GROUP Ltd (2024) Expanded polystyrene (EPS) and the environment. Available from: https://www.eps.co.uk/pdfs/eps_and_the_environment.pdf
4. Gausepohl H, Nießner N (2001) Polystyrene and styrene copolymers. In: Buschow KHJ, Cahn RW, Flemings MC, Ilschner B, Kramer EJ, Mahajan S, et al. (eds) Encyclopedia of materials: science and technology. Elsevier, Oxford, p 7735–7741. Available from: https://www.sciencedirect.com/science/article/pii/B0080431526013899

5. Raps D, Hossieny N, Park CB, Altstädt V (2014) Past and present developments in polymer bead foams and bead foaming technology. Polymer (Guildf).56:5–19. Available from: https://www.sciencedirect.com/science/article/pii/S003238611401012X
6. British Plastics Federation EPS GROUP (2024) Sustainability & expanded polystyrene. Available from: https://www.eps.co.uk/sustainability/index.html
7. Acierno S, Carotenuto C, Pecce M (2009) Compressive and thermal properties of recycled EPS foams. Polym Plast Technol Eng. 49(1):13–19. https://doi.org/10.1080/03602550903282994
8. Uttaravalli AN, Dinda S, Gidla BR (2020) Scientific and engineering aspects of potential applications of post-consumer (waste) expanded polystyrene: a review. Process Safety Environ Protection 137:140–148. Available from: https://www.sciencedirect.com/science/article/pii/S0957582020301154
9. Wypych G (2022) Handbook of polymers. Elsevier
10. International Union of Pure and Applied Chemistry (IUPAC) (2014) Front Matter. In: Nomenclature of organic chemistry. Cambridge: royal society of chemistry, pp P001–P004
11. Speight JG (2020) Chapter 14—monomers, polymers, and plastics. In: Speight JG (ed) Handbook of industrial hydrocarbon processes (second edition). Gulf Professional Publishing, Boston, pp 597–649. Available from: https://www.sciencedirect.com/science/article/pii/B978012809923000014X
12. Polymer Science Learning Center. Polystyrene (2023) [cited 31 Aug 2024] Available from: https://pslc.ws/macrog/styrene.htm
13. Gil-Jasso ND, Giles-Mazón EA, Soriano-Giles G, Reinheimer EW, Varela-Guerrero V, Ballesteros-Rivas MF (2022) A methodology for recycling waste expanded polystyrene using flower essential oils. Fuel 307:121835
14. Noguchi T, Tomita H, Satake K, Watanabe H (1998) A new recycling system for expanded polystyrene using a natural solvent. Part 3. Life cycle assessment. Packaging Technology and Science 11(1):39–44
15. Construcción Fácil. Cómo hacer una capa de poliestireno expandido [Video] (2023) [cited 31 Aug 2024]. Cómo hacer una capa de poliestireno expandido [Video]. Available from: https://www.youtube.com/watch?v=HhlW7Kn7U9U&t=213s
16. Rossacci J, Shivkumar S (2003) Bead fusion in polystyrene foams. J Mater Sci 38:201–206
17. Zhai W, Kim YW, Jung DW, Park CB (2011) Steam-Chest molding of expanded polypropylene foams. 2. Mechanism of interbead bonding. Ind Eng Chem Res 50(9):5523–5531. https://doi.org/10.1021/ie101753w
18. Malewska E, Prociak A, Świdzińska-Grela N, Kurańska M (2024) The polyurethane-polystyrene composite—influence of the blowing agent type on the foaming process, the structure and the properties. J Composites Sci 8(4):135
19. National polystyrene systems (2011) Technical data information- rigid cellular expanded polystyrene [cited 31 Aug 2024] pp 1–12. Available from: https://nationalpolystyrene.com.au/wp-content/uploads/2020/02/Technical-Data2-National-Poly-10022020.pdf
20. International organization for standardization (2017) ISO 7071:2017 Plastics—expanded polystyrene for thermal insulation—specification. Geneva, Switzerland
21. ASTM International (2019) ASTM C578–19 standard specification for rigid, cellular polystyrene thermal insulation. West Conshohocken, PA
22. National Polystyrene Systems (2021) Rigid Celullar Expanded Polystyrene—Technical Data Information [cited 31 Aug 2024]. Available from: https://nationalpolystyrene.com.au/wp-content/uploads/2021/02/Technical-Data-Information-28.01.2021.pdf
23. Future Foams. EPS manual (2023) [cited 31 Aug 2024]. Available from: https://www.futurefoams.com.au/wp-content/uploads/2023/06/Future-Foams-EPS-Manual.pdf
24. ARPADIS. BLOWING AGENT [cited 31 Aug 2024] BLOWING AGENT. Available from: https://www.arpadis.com/blowing-agent/
25. Zhu Z, Zong JH, Park CB, Choudhary M (2009) Effect of loss of blowing agents on thermal insulation properties of polystyrene foams
26. The Dow Chemical Company. Process for preparing extruded polystyrene foams with enhanced thermal insulating properties (1993). United States Patent and Trademark Office [cited 31 Aug 2024]. Available from: https://patents.google.com/patent/US5250577A/en

27. Sun Y, Ueda Y, Suganaga H, Haruki M, Kihara S ichi, Takishima S (2016) Experimental and simulation study of the physical foaming process using high-pressure CO_2. J Supercrit Fluids 107:733–745. Available from: https://www.sciencedirect.com/science/article/pii/S0896844615300929
28. Ray SS, Banerjee R (2022) Chapter 1—introduction to polymer foams and foaming. In: Ray SS, Banerjee R (eds) Foamability of thermoplastic polymeric materials. Elsevier. pp 1–16. Available from: https://www.sciencedirect.com/science/article/pii/B9780323907675000030
29. Bridson JH, Gaugler EC, Smith DA, Northcott GL, Gaw S (2021) Leaching and extraction of additives from plastic pollution to inform environmental risk: a multidisciplinary review of analytical approaches. J Hazard Mater 414:125571. Available from: https://www.sciencedirect.com/science/article/pii/S0304389421005343
30. De-la-Torre GE, Dioses-Salinas DC, Pizarro-Ortega CI, Saldaña-Serrano M (2020) Global distribution of two polystyrene-derived contaminants in the marine environment: a review. Mar Pollut Bull 161:111729. Available from: https://www.sciencedirect.com/science/article/pii/S0025326X2030847X
31. Li H, Zhang Z, Sun Y, Wang W, Xie J, Xie C, et al (2021) Tetrabromobisphenol A and hexabromocyclododecanes in sediments and biota from two typical mangrove wetlands of South China: Distribution, bioaccumulation and biomagnification. Sci Total Environ 750:141695. Available from: https://www.sciencedirect.com/science/article/pii/S0048969720352244
32. Agarwal S, Gupta RK (2017) 29—Plastics in buildings and construction. In: Kutz M, (ed) Applied plastics engineering handbook (2nd Edition). William Andrew Publishing, pp 635–649. Available from: https://www.sciencedirect.com/science/article/pii/B9780323390408000304
33. Zhu SE, Yang WJ, Zhou Y, Pan WH, Wei CX, Yuen ACY, et al (2022) Synthesis of zinc porphyrin complex for improving mechanical, UV-resistance, thermal stability and fire safety properties of polystyrene. Chem Eng J 442:136367. Available from: https://www.sciencedirect.com/science/article/pii/S1385894722018629
34. Pushpadass HA, Weber RW, Hanna MA (2008) Expansion, morphological, and mechanical properties of starch–polystyrene foams containing various additives. Ind Eng Chem Res 47(14):4736–4742. https://doi.org/10.1021/ie071049h
35. Varnagiris S, Girdzevicius D, Urbonavicius M, Milcius D (2017) Incorporation of SiO_2 and TiO_2 additives into expanded polystyrene foam using physical vapour deposition technique. Energy Procedia 128:525–532. Available from: https://www.sciencedirect.com/science/article/pii/S1876610217339176
36. Varnagiris S, Urbonavicius M, Tuckute S, Lelis M, Milcius D (2017) Development of photocatalytically active TiO_2 thin films on expanded polystyrene foam using magnetron sputtering. Vacuum 143:28–35. Available from: https://www.sciencedirect.com/science/article/pii/S0042207X17302166
37. Waldo-Mendoza MA, Martínez-Jothar L, Oros-Ruiz S, Villabona-Leal EG, Céspedes-Covarrubias LM, Quiñones-Jurado ZV, et al (2016) Antibacterial additive for polystyrene based on silver nanoparticles supported on titanium dioxide. Int J Polym Sci 2016(1):1641352. https://doi.org/10.1155/2016/1641352
38. García MT, Gracia I, Duque G, Lucas A de, Rodríguez JF (2009) Study of the solubility and stability of polystyrene wastes in a dissolution recycling process. Waste Management 29(6):1814–1818. Available from: https://www.sciencedirect.com/science/article/pii/S0956053X09000038
39. García MT, Duque G, Gracia I, de Lucas A, Rodríguez JF (2009) Recycling extruded polystyrene by dissolution with suitable solvents. J Mater Cycles Waste Manag 11(1):2–5. https://doi.org/10.1007/s10163-008-0210-8
40. Isowall Group. EPS properties ideal for food packaging [cited 31 Aug 2024]. EPS properties ideal for food packaging. Available from: https://isowall.co.za/eps-properties-ideal-for-food-packaging/

"Characteristics of Expanded Polystyrene Foams"

Dafne L. Ortega-Solis, Victor Varela-Guerrero, and María F. Ballesteros-Rivas

Abstract Expanded polystyrene, also known as EPS, is defined as a cellular and rigid plastic material composed mostly of air, which is characterized by its closed cellular structure. This structure gives it a series of exceptional properties, such as excellent thermal and acoustic insulation, lightness, strength and durability. Due to these properties of the material, it becomes attractive for applications in construction, packaging and, in addition, being recyclable, makes it an increasingly attractive option in a context of growing concern for the environment.

Keywords Expanded polystyrene (EPS) · EPS density · Physical properties · Chemical properties · Biological properties of eps

1 Introduction

Polystyrene is a thermoplastic material defined as a hard, rigid and shiny synthetic resin produced by the polymerization of styrene. Due to its unique combination of lightness and strength, it is widely used in the construction, food, etc. industries [1]. There are several types of polystyrene, including general purpose polystyrene (GPPS), high impact polystyrene (HIPS) and expandable polystyrene (EPS). In Mexico, the national production of polystyrene reaches 417 thousand tons of installed capacity and the main producers are Styropek (12%), Novidesa (12%), Resirene (36%), and BASF (40%) (Graph 1).

D. L. Ortega-Solis · V. Varela-Guerrero (✉) · M. F. Ballesteros-Rivas (✉)
Facultad de Química, Universidad Autónoma del Estado de México, Toluca de Lerdo, México
e-mail: vvarelag@uaemex.mx

M. F. Ballesteros-Rivas
e-mail: fballesterosr@uaemex.mx

Centro Conjunto de Investigación en Química Sustentable, UAEM-UNAM, Toluca de Lerdo, Mexico

A. Khan et al. (eds.), *Uses and Products of Recycled Expanded Polystyrene Foam Wastes*, Sustainable Materials and Technology,
https://doi.org/10.1007/978-981-96-5618-9_6

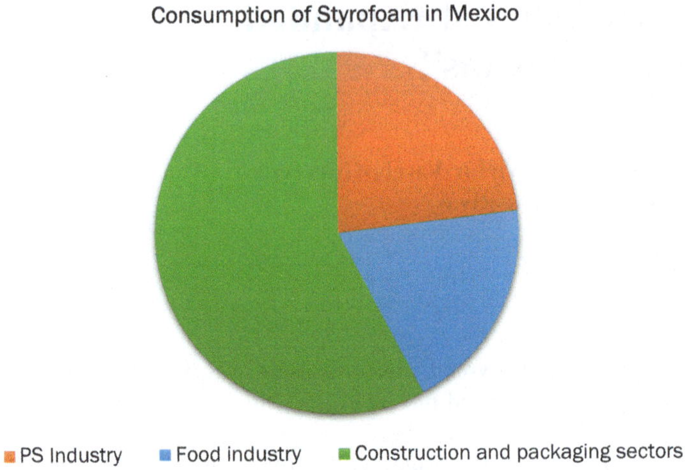

Graph 1 Consumption of Styrofoam in Mexico. *Source* National Association of Plastic Industries (ANIPAC)

According to the National Association of Plastic Industries (ANIPAC), consumption of Styrofoam in Mexico is 125 thousand tons per year, which represents 29.97% of the total PS industry. Of these, 25% is used to manufacture disposable products for the food industry and the remaining 75% is divided between the construction and packaging sectors [2].

The applications of polymeric foams are always based on their properties. Polystyrene foam (EPS) is an industrial material widely used in road construction, automobiles, architecture, etc. [3] and is defined as a hard, rigid, shiny synthetic resin produced by the polymerization of styrene and is divided into two types: solid polystyrene (solid PS) and expanded polystyrene (EPS). The versatility of polystyrene (PS) lies in its ability to be reprocessed into resin, allowing it to be reused in a wide range of applications such as food packaging, office equipment, etc. [4]. By combining properties such as shock absorption, low cost, and simple production, EPS is positioned as an ideal material to protect products during transport and as a protective element in construction. Expanded polystyrene (EPS) is a rigid, lightweight foam known for its excellent insulating and shock-absorbing properties. It is produced from styrene and pentane, compounds derived from petroleum and natural gas. The process involves synthetic beads that, when heated, expand and form a closed-cell foam. These beads contain a foaming agent such as pentane, which allows the density and final texture of the product to be controlled [5] (Fig. 1) [1].

These components combine to create a material with exceptional insulation and shock absorption capabilities, making it a versatile material with numerous applications. Its final composition is mainly air (98%) and a small proportion of raw materials (2%), which contributes to its exceptional performance. The conversion of expanded EPS takes place in three stages (Figs. 2and 3).

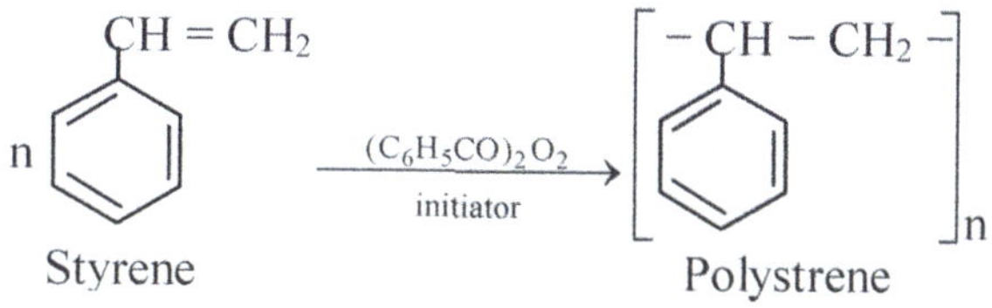

Fig. 1 Expanded polystyrene molecule

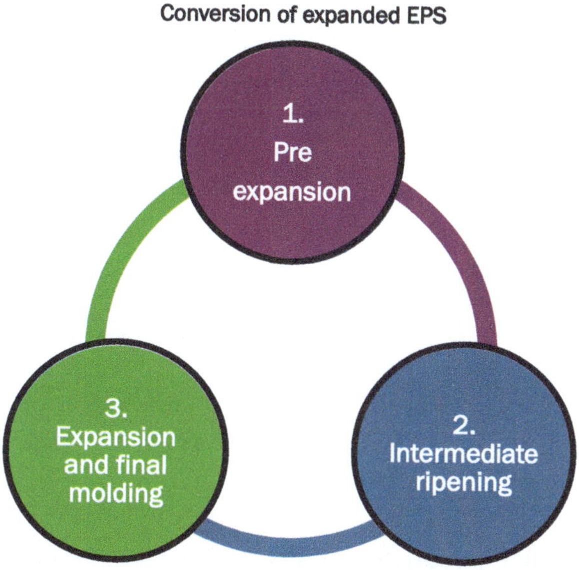

Fig. 2 Conversion of expanded EPS

Polystyrene production begins with the extraction of crude oil, from which naphtha is obtained through a distillation process. Naphtha, after being subjected to desulfurization, is transformed into benzene and ethylene, precursors of styrene. This monomer, through a polymerization process, is transformed into long molecular chains that give rise to polystyrene. This polymer, in the form of small beads, is pre-expanded, matured and stabilized, and expanded in the mold to obtain a wide variety of products, such as EPS containers [6].

During the EPS synthesis process, it can be modified with additives. These added substances make it possible to modify or improve the characteristics of EPS. The additives, which can be organic or inorganic in nature, are incorporated in minimal proportions to modify the properties of expanded polystyrene [7]. The additives used in the manufacture of EPS are classified into several types, each with a specific function:

- **Plasticizing additives**: Increase the flexibility and malleability of EPS, making it more resistant to deformation.
- **Foaming additives**: Produce air bubbles and this reduces the weight of the material, improving the thermal and acoustic insulation capacity.

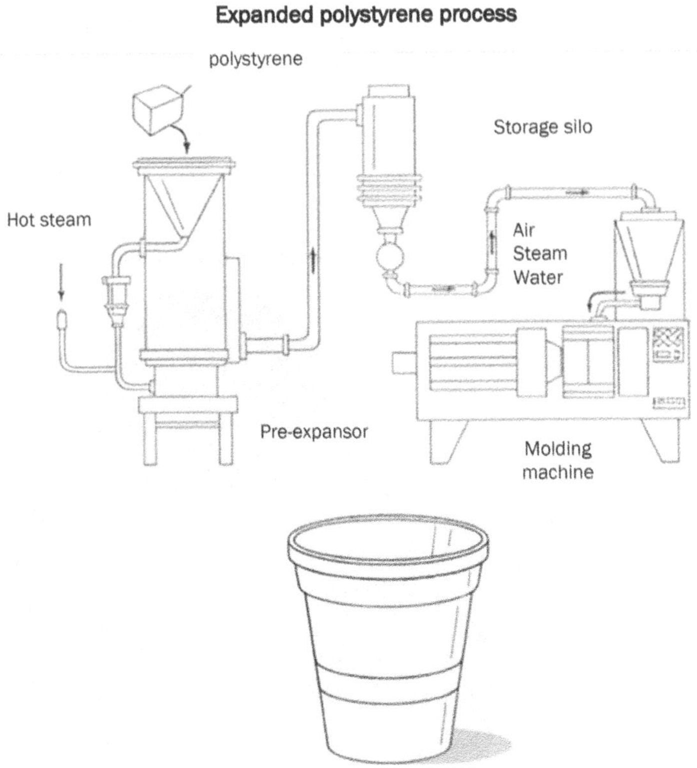

Fig. 3 Expanded polystyrene process. *Source* How products are made?

- **Reinforcing agents**: Increase the rigidity and impact resistance of EPS, improving its mechanical properties.
- **Antioxidant additives**: Protect EPS from degradation when exposed to light and heat.
- **Slip additives**: Improve flow capacity during the molding process. These additives are especially useful in the production of large parts or parts with complex geometries.

Among the most notable characteristics of expanded polystyrene foams are those shown in Fig. 4. In addition, the natural color of EPS is white due to the refraction of light.

As we have already mentioned, EPS is composed of a greater amount of air, which makes it a much lighter packaging material, reducing the total weight of the packaging and thus saving on the impact of transportation for these producers, as it saves on fuel consumption. Its ease of cutting and molding makes it a blank canvas for creativity, allowing the manufacture of customized parts with great precision. Whether in construction or product packaging, EPS adapts to any project that requires unique and complex shapes, offering an economical and versatile solution [8].

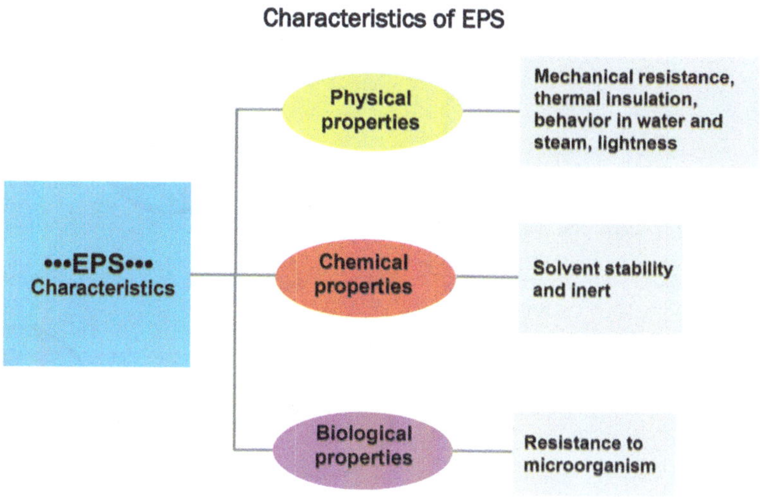

Fig. 4 Characteristics of EPS [8]

2 Physical Properties

2.1 Density

One of the main characteristics of EPS is density, defined as the mass per unit volume, a key property that directly influences its performance and applications. Higher density implies a more compact structure, which translates into greater mechanical strength, but can also reduce its insulating properties. The density of EPS foams can vary depending on the synthesis process, but these densities range from around 10 kg/m^3 to over 40 kg/m^3.

During the synthesis process of EPS, the density of the material can be determined, and we can determine these between three: low, medium and high density (Fig. 5).

- **Low density EPS**: This process involves the expansion of EPS beads by steam with a foaming agent. Its density value is between 10 and 20 kg/m^3. Its applications are in sectors where it needs to be used as a lightweight material with a high thermal insulation capacity, as packaging for the protection of fragile products, in flotation for the fishing and aquaculture industry and in construction to be used as a thermal insulation material.
- **Medium density EPS**: In this process, the pre-expanded beads are subjected to higher pressures and longer molding times, which allow for molding and creating a denser EPS foam structure. The density range is between 20 and 30 kg/m^3.
- **High density EPS**: Molding is done in multiple stages and cooling must be rapid to achieve a more compact EPS foam structure. This process is in line with the previous ones, the variation here is that the steam conditions and mold

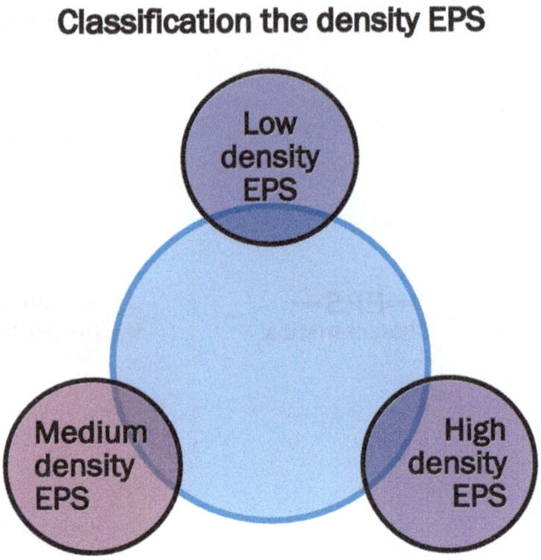

Fig. 5 Classification the density EPS

temperatures are much more controlled. The density of this type of polymer is between 30 and 40 kg/m^3. This foam stands out for its three properties, and offers a valuable alternative for applications that require these three properties:

- **Greater strength**: By packing more material into the same volume, the result is a stronger and more rigid foam that is applied to materials that require greater load capacity, durability and resistance.
- **Greater durability**: It does not deform or break, which allows it to be used in applications that need to withstand greater impacts and compression forces.
- **Greater resistance**: Used in applications that are in contact with water or humidity.

The density of expanded polystyrene foams is a determining factor in their performance. This physical property directly influences characteristics, making EPS ideal for a wide range of applications.

2.2 *Mechanical Strength*

The density of the material is closely correlated with the mechanical strength properties. The closed cell structure of EPS does not allow for a conventional tensile test, this closed structure gives the material a limited capacity for deformation under tension, which restricts its application in situations requiring stretching. However, in practice, EPS is mainly used under compressive loading conditions. By establishing a deformation criterion of 10%, a reference value is obtained that facilitates comparison between different types of EPS and ensures compliance with quality standards. In summary, we can establish 4 properties that determine the mechanical strength of EPS (Fig. 6).

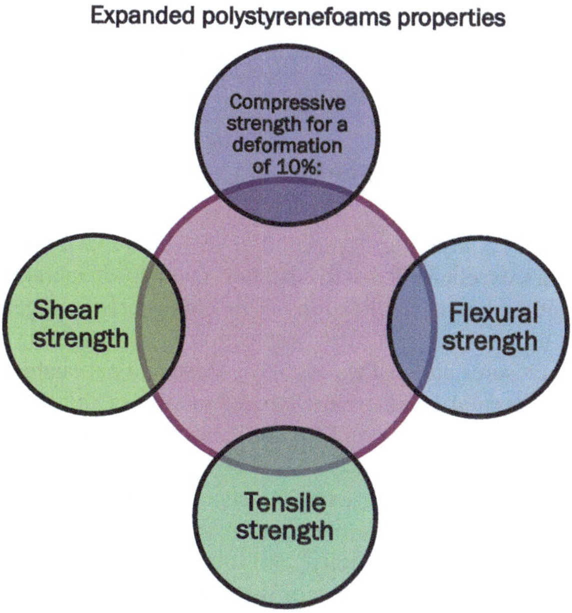

Fig. 6 EPS properties that determine mechanical strength

Expanded polystyrene particles have specific characteristics such as having low thermal conductivity and being resistant to low temperatures (mentioned later), for this reason this material is used in soil science, one of its main applications is to mix and compact it to form a new type of renewable environmental fill material. The very light density of renewable EPS gives a value of less than 1.2 g/cm^3, which is reflected in a positive change for environmental pollution by decreasing its volume in addition to solving the problems of freezing–thawing and soil settlement.

The physical and mechanical properties of soil in its freezing and thawing cycles have impacted so much that many scientists have studied how to improve this deterioration that occurs in the soil. With the characteristics that EPS presents, this material arises as a new opportunity to be widely used in soft soil foundations, road filling, wall filling, pipe filling and other projects. Authors such as: Qishan Jiang et al. studied the soft soil subgrade treatment technologies, concluding that replacing lightweight soil is fast, reliable and economical, and established a calculation method for the thickness of lightweight soil replacement [9].

Li et al. conducted an experiment on filling the retaining wall with lightweight soil with EPS particles and the results showed that it could greatly reduce the soil pressure and vertical settlement. They analyzed the decompression mechanism and settlement deformation mechanism of lightweight soil fills in retaining wall engineering [10].

Another of the most recent researches demonstrated the use of EPS particles as soil filler. The authors Lifang et al. proposed a scheme of lightweight soil filler with expanded polystyrene (EPS) particles in cold regions [11] in their research they conclude that EPS particles allow to improve the frost resistance of light soils, however, when it increases above 2%, the soil cement bound to the EPS particles is limited and the frost resistance of light soils decreases [11].

2.3 Thermal Insulation

The EPS structure, composed of 98% air trapped inside the cellular structure, makes it an excellent thermal insulator, capable of maintaining stable temperatures in a wide range of applications. This characteristic, together with its resistance, makes it an ideal material for construction as an insulating material for different building enclosures and/or for packaging fresh and perishable foods [12].

Thermal conductivity λ is a function of the thermal insulation capacity of the material, in the case of EPS this value varies. The graph shows this influence (Graph 2) [12].

In fact, current research has succeeded in developing EPS with even lower thermal conductivity, thanks to innovative raw materials incorporating special additives [13]. These "low conductivity" or "infrared absorbing" materials, usually gray in color, further expand the range of possibilities of EPS in applications where thermal insulation is critical [12]. In addition to dimensional changes caused by temperature fluctuations, there are other alterations that expanded polystyrene can undergo when exposed to the effects of thermal action. This material can be used at extremely low temperatures without its properties being affected, except for dimensional variations due to contraction. However, its upper temperature limit is around 100 °C for

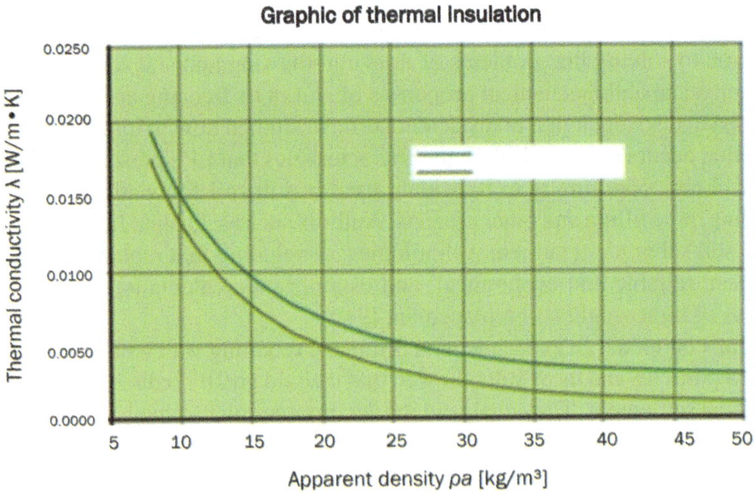

Graph 2 Graphic of thermal insulation, general values vary from 0.043 to 0.029 for all applications. *Source* Asociación Nacional de poliestireno expandido (ANAPE)

short-term use and 80 °C for continuous use under a load of 20 kPa. The thermal conductivity values have been reported by BSL/Dow in the graph.

ASTM C578 is an international specification that establishes the requirements and characteristics of expanded polystyrene thermal insulation panels. This standard evaluates properties such as thermal conductivity, compressive strength and water absorption, thus ensuring optimal performance in construction applications.

2.4 Compression Resistance

EPS is the ideal solution for packaging large objects. Its ability to absorb impacts makes it an exceptional shock absorber, while its low density makes it easy to handle and store. In addition, its cellular structure gives it great resistance to compression, allowing loads to be stacked without risk of deformation [12].

2.5 Water Absorption

The closed cellular structure of EPS acts as a barrier against liquid penetration. This property is more present in high-density EPS, although it does show a slight absorption when immersed in water. These values range between 1 and 3% by volume, which is why EPS foams are considered non-hygroscopic [12].

2.6 Water Vapor Diffusion

On the other hand, the ability of EPS to resist the passage of water vapor is a key factor in its application. The μ factor, which compares the resistance of EPS to that of air, is used to determine the behavior of the material in relation to moisture. Water vapor can penetrate the cellular structure of EPS and this permeability is influenced by the differences in pressure and temperature on both sides of the material. The resistance of EPS to water vapor diffusion is quantified by the μ factor, which compares its resistance to that of air. In the case of EPS, this value usually ranges between 20 and 100, indicating a higher resistance to water vapor diffusion compared to air. The low moisture absorption and excellent dimensional stability of EPS, over 98%, make it an ideal packaging material for sensitive products [14].

2.7 Dimensional Stability

The evaluation of this property is determined by the coefficient of thermal expansion and is situated within the values that range from 5–7 × 10–5 K^{-1}, that is, between 0.05 and 0.07 mm per meter of length and degree Kelvin and these variations are caused by the action of heat or subsequent contraction of the material [14].

2.8 Temperature Stability

Expanded polystyrene foams can be altered by thermal action. This material can be used safely at extremely low temperatures, without affecting its properties, except for dimensional variations caused by contraction. With regard to the upper end, the temperature limit for use is around 100 °C for short-term actions, and around 80ºC for continuous actions and with the material subjected to a load of 20 kPa.

2.9 Impact Absorption

The high energy absorption capacity of EPS makes it the material of choice for protecting delicate products during handling and storage. This characteristic makes it indispensable in critical applications such as safety helmets, sports equipment and child car seats, where protection is paramount [15].

3 Chemical Properties

EPS is a material that is resistant to many common chemicals and solvents, which contributes to its durability and longevity and continued use in a variety of applications. The following table details its chemical stability to some common solvents [12].

Active substance	Stability
Saline solution	Stable
Soaps and surfactant solutions	Stable
Chlorine	Stable
Dilute acids	Stable
Hydrochloric acid (35%), nitric acid (50%)	Stable
100% concentrated acids	Not stable
Alkaline solutions	Stable

(continued)

(continued)

Active substance	Stability
Organic solvents (acetone, esters)	Not stable
Saturated aliphatic hydrocarbons	Not stable
Paraffin oils, Vaseline	Stable
Diesel oil	Not stable
Alcohols (methanol, ethanol)	Estable

4 Biological Properties

4.1 Microbiological Resistance

All the characteristics of EPS allow it to be an inherently resistant material to the action of microorganisms such as bacteria and fungi, and this makes EPS products meet health, safety and hygiene requirements and can be used as packaging products intended for the food area. However, the disadvantage it presents is that when EPS is in the presence of dirt, it can be a carrier of microorganisms, without participating in the biological process [8].

Although EPS offers numerous beneficial properties, its life cycle reveals a considerable environmental impact. The presence of contaminated EPS microplastics represents a serious risk to the health of ecosystems and living organisms. From its manufacture to its final disposal, this material contributes significantly to the pollution of ecosystems, positioning it as a global environmental challenge. It was mentioned that the density of EPS is much lower and this environmental behavior is of importance for the marine environment (Turner 2020). Microplastics have emerged as a potential carrier of many pollutants, such as persistent organic pollutants [16], antibiotics, and pathogens [17].

5 EPS Applications According to Its Characteristics

5.1 Building and Construction

EPS is extremely versatile and offers the perfect combination of mechanical and physical properties, including dimensional stability, permanent R-value, inherent water resistance and compressive strength, as well as offering longevity and sustainable green credentials [18].

Efficient, durable and time-tested for over 60 years, EPS insulation is an effective way to meet design specifications and climate protection objectives that allow it to be used in construction. In this area we can highlight three characteristic applications:

- **EPS Building Systems**

Concrete, a mixture of cement, sand and gravel, is a widely used building material. Insulated concrete form (ICF) panels are an innovative alternative [18] and are defined as structural wall panels made of a concrete core poured into interlocking expanded polystyrene (EPS) that holds the concrete together during the curing operation [19].

These panels, composed of a reinforced concrete core within an expanded polystyrene mould, offer an efficient construction system with excellent insulating properties. The EPS remains permanently in place as part of a wall panel and thus provides thermal insulation to the building and the reinforced concrete provides the structural system to the construction [20], that is, it is a three-layer composite material that is joined by joining a thin layer (skin) on each side of a thick layer (core), it is the core that is made of EPS sandwiched between two oriented strand boards [21].

Evaluation of mechanical properties of EPS is critical given its predominant volume in ICFs. Few types of experimental research have been reported on exploring the characteristics of EPS. Chen et al. [Static and dynamic mechanical properties of expanded polystyrene] studied the compression and tensile characteristics of EPS with density of 13.5 kg/m^3 and 28 kg/m^3. Beju et al. [22] reported the results of compression, flexural and water absorption tests of EPS with densities of 12, 15, 20 kg/m^3 and concluded that density is the main governing factor to control the properties of EPS.

- **Geofoam for Civil Engineering**

EPS geofoam is a versatile material with multiple applications in civil engineering and construction, it is primarily composed of expanded polystyrene (EPS), a lightweight closed-cell foam material. This composition involves the expansion of polystyrene beads, which are then fused together to form large blocks or panels. Thanks to its low density, high compressive strength and stability, this material offers innovative solutions to a wide range of challenges, from construction on soft ground to earthquake protection. Its ability to dampen vibrations and noise makes it an ideal choice for projects in urban areas, it is used in the construction of roads, highways and airfields, railway track systems, beneath refrigerated storage buildings, sports stadiums and storage tanks to prevent freezing and ground heaving, and in underground building segments to reduce seasonal heating and cooling requirements [23].

As an engineered product, EPS geofoam allows builders to design for key geosynthetic functions and select the best combination of products to achieve project objectives. The relevant specification for EPS geofoam is ASTM D6817, Standard Specification for Rigid Cellular Polystyrene Geofoam.

- **Bridge abutments**

EPS geofoam complements the properties of expanded polystyrene such as high compressive strength, allowing it to be used to replace compressible soils or in place of heavy fill materials and thus safely support the load of the roadway without overloading the underlying soils. This generally results in less differential movement at the interface between the bridge and the approach fill, reducing the cost of constructing the approach slab and its long-term maintenance.

- **Stadium seating**

EPS geofoam can be used to form seating in places such as auditoriums, cinemas, gymnasiums and churches. The high compressive strength and light weight of EPS geofoam make it ideal for both new construction and renovation projects.

5.2 Packaging—EPS Packaging for Use in Food

It was previously mentioned that another of the characteristics of EPS is its zero capacity to absorb water and humidity, fundamental properties for use in food packaging and to preserve any fresh food, such as meat products, fish, seafood, fruits and vegetables, in addition to not being a carrier of fungal and bacterial cultures due to its biological properties. In addition to these characteristics, EPS has the property of absorbing impacts and being a thermal insulator and these capacities allow the food not to suffer impacts during transport in addition to keeping the food in optimal conditions. EPS packaging can be in direct contact with food because it complies with the requirements of international health regulations [24].

6 Sustainability—Characteristics of EPS in the Environment

The production of polystyrene has a considerable environmental impact. From the extraction of raw materials, such as oil, to the transformation processes and final disposal of the product, greenhouse gas emissions are generated that contribute to climate change.

However, modern polystyrene manufacturing, despite being an alternative to harmful CFCs, presents a new problem: it releases significant quantities of hydrofluorocarbons (HFCs), potent greenhouse gases. Initially adopted as a less harmful solution, HFCs have proven to be an even greater environmental problem. Faced with this situation, governments have implemented measures to drastically reduce their emissions (Graph 3).

The range of properties of expanded polystyrene is very wide and highly customizable and depends on various factors such as the manufacturing method, the gas used,

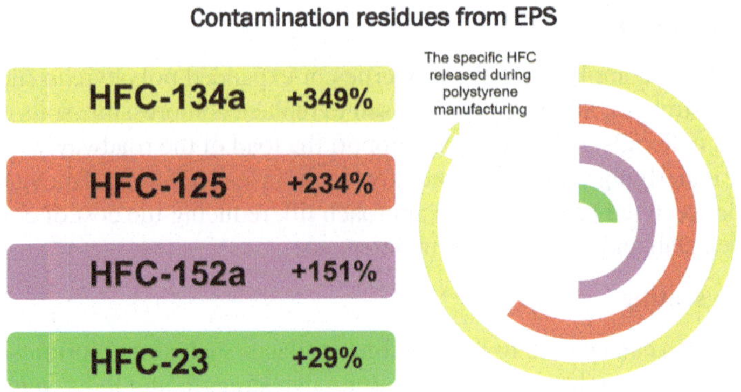

Graph 3 Contamination residues from EPS

the cellular structure and the base material [3]. For example, the density of EPS is a critical factor in determining its properties. According to the British Plastics Federation, the density of EPS ranges from 12 to 50 kg/m^3, which directly influences its compressive strength. Lower density equates to higher strength, allowing EPS to be adapted to a variety of applications [7].

7 Conclusion

In this chapter we have identified expanded polystyrene as a versatile material due to its excellent chemical, physical and biological properties, highlighting thermal insulation. The density of EPS turns out to be a determining factor in its characteristics and its various applications. Some examples of how EPS acts in the construction and packaging industry were shown. We observed that as the density increases, the resistance of the material improves but one of its main characteristics is reduced; its insulation capacity. We can close the chapter by concluding that depending on the application, the desiccated density plays an important role.

References

1. Mexico PT (2024) ¿Qué es el poliestireno (PS)? tipos y aplicaciones desglosadas. Available from: https://www.pt-mexico.com/articulos/que-es-el-poliestireno-ps-tipos-y-aplicaciones
2. Rennueva. La startup mexicana de la UNAM que recicla unicel después de consumo, es pionera en el mundo. 2024; Available from: https://www.travelbymexico.com/blog/19392-rennueva-la-startup-mexicana-de-la-unam-que-recicla-unicel-despues-de-consumo-es-pionera-en-el-mundo/
3. Andena L et al (2019) Compression of polystyrene and polypropylene foams for energy absorption applications: a combined mechanical and microstructural study. J Cell Plast 55(1):49–72

4. Banger A et al (2023) Polymeric foams: mechanisms and properties. Polymeric foams: fundamentals and types of foams (Volume 1). American Chemical Society, pp 43–61
5. Cui C et al (2016) Stress–strain relationship in axial compression for EPS concrete. Constr Build Mater 105:377–383
6. Enarevba DR, Haapala KR (2023) A comparative life cycle assessment of expanded polystyrene and mycelium packaging box inserts. Procedia CIRP 116:654–659
7. Omnexus (2024) Expanded polystyrene (EPS): ultimate guide on foam insulation material. Available from: https://omnexus.specialchem.com/selection-guide/expanded-polystyrene-eps-foam-insulation
8. aispor. Poliestireno expandido: qué es, propiedades y usos (2024) Available from: https://aispor.es/blog/propiedades-del-poliestireno-expandido/
9. Jiang QC, Wang J, Liu J (2019) Method for determining the thickness of light soil in embankment replacement on soft ground. Highway 64(04):84–88
10. Li TH, Zhang D, Chen F, Li Z (2021) Experimental study on long term service performance of heavy-haul railway light-weight foamed concrete in bridge-subgrade transition section. Railway 61(07):80–84
11. Mei L et al (2023) Physical and mechanical properties of expanded polystyrene (EPS) particle lightweight soil under freeze-thaw cycles. ACS Omega 8(34):31365–31372
12. anape. ¿Qué es el EPS? (2024) Available from: https://anape.es/producto/
13. Expanded Polystyrene (EPS): Ultimate Guide on Foam Insulation Material. Available from: https://omnexus.specialchem.com/selection-guide/expanded-polystyrene-eps-foam-insulation
14. Modopack. Especificaciones Tecnicas poliestireno (2020)
15. Di Landro L, Sala G, Olivieri D (2002) Deformation mechanisms and energy absorption of polystyrene foams for protective helmets. Polym Testing 21(2):217–228
16. de Sá LC et al (2018) Studies of the effects of microplastics on aquatic organisms: What do we know and where should we focus our efforts in the future? Sci Total Environ 645:1029–1039
17. Bridson JH et al (2021) Leaching and extraction of additives from plastic pollution to inform environmental risk: a multidisciplinary review of analytical approaches. J Hazard Mater 414:125571
18. Alliance I (2024) EPS building systems. Available from: https://www.epsindustry.org/eps-building-systems
19. Arun Solomon A, Hemalatha G (2020) Characteristics of expanded polystyrene (EPS) and its impact on mechanical and thermal performance of insulated concrete form (ICF) system. Structures 23:204–213
20. Dusicka P, Kay T (2011) In-Plane lateral cyclic behavior of insulated concrete form grid walls. J Struct Eng 137(10):1075–1084
21. Sulong H, Mustapa A, Abdul Rashid MK (2019) Application of expanded polystyrene (EPS) in buildings and constructions: a review. J Appl Polym Sci 136
22. Beju YZ, Mandal JN (2017) Expanded polystyrene (EPS) geofoam: preliminary characteristic evaluation. Procedia Engineering 189:239–246
23. Alliance I (2024) EPS Geofoam: effective civil engineering solutions for a broad range of applications. Available from: https://www.epsindustry.org/geofoam
24. TFT, El EPS es un material con alto potencial para el mercado de empaques (2022)

The manufacturer's authorised representative in the EU is Springer Nature Customer Service Centre GmbH, Europaplatz 3, 69115 Heidelberg, Germany. If you have any concerns regarding our products, please contact ProductSafety@springernature.com

Printed and bound by CPI Group (UK) Ltd, Croydon, CR0 4YY
01/10/2026
02225916-0004